NO LONGER PROPERTY OF
SEATTLE PUBLIC LIBRARY

Transient

and

Strange

Transient

and

Strange

Notes on the Science of Life

Nell Greenfieldboyce

W. W. NORTON & COMPANY
Independent Publishers Since 1923

Copyright © 2024 by Nell Greenfieldboyce

All rights reserved
Printed in the United States of America
First Edition

For information about permission to reproduce selections from this book, write to
Permissions, W. W. Norton & Company, Inc., 500 Fifth Avenue, New York, NY 10110

For information about special discounts for bulk purchases, please contact
W. W. Norton Special Sales at specialsales@wwnorton.com or 800-233-4830

Manufacturing by Lakeside Book Company
Book design by Beth Steidle
Production manager: Lauren Abbate

ISBN: 978-0-393-88234-6

W. W. Norton & Company, Inc.
500 Fifth Avenue, New York, N.Y. 10110
www.wwnorton.com

W. W. Norton & Company Ltd.
15 Carlisle Street, London W1D 3BS

1 2 3 4 5 6 7 8 9 0

Contents

Introduction

For nearly thirty years, I have made my living by writing about science. Much of that time has been spent working for National Public Radio (NPR). I usually produce two-to-eight-minute audio stories that are designed to inform or delight: neat, tidy sonic tales that convey news about, say, the color of dinosaur eggs, or possible signs of life on Venus, or sharks that swim for centuries beneath the Arctic ice. A workday might involve chatting with a researcher who studies carnivorous plants or flying snakes or gravitational waves that roll through the fabric of space-time. To report my stories, I've climbed into one of NASA's space shuttles (on the ground, thankfully) and onto a boat that was hunting for a Civil War–era submarine. I've gone down into a coal mine, where deafening machines ripped black rock from the Earth's crust, and into the control room of a particle collider, where magnets accelerated beams of tiny particles to almost the speed of light, smashing them together so that scientists could pick through the subatomic rubble.

When I became a parent, I suddenly had a new audience. A much smaller audience—only two, instead of millions—but with higher emotional stakes. With my son and daughter, I generally tried to err on the side of staying silent instead of sharing a thought or observation or memorized fact. Whatever I might say seemed unnecessary and often unequal to their own powers of perception. The child-as-scientist has become a cliché, but I marveled at my children's relentless exploration, their intense focus, their ruthless experimentation, their admirable lack of preconceptions. One morning, for example, when my son was about two and a half years old, I woke up and came downstairs to find the refrigerator door open and my son sitting inside the fridge, on an interior ledge, facing the shelves full of condiments. "Are you finding anything interesting in there?" I asked (ignoring the panic synapses that clanged *child in a refrigerator, danger! danger!*). He said, "I'm trying to open this. What is this?" He held up a stick of butter. I took it, partially unwrapped it, and handed it back. He bit into it thoughtfully, then dug his little fingers into it and pulled off a greasy lump, saying, "Here is some butter for you." I was touched by this gesture, his desire to include me in the moment of discovery.

What I do tell them often proves useless, and unconvincing. When my toddling daughter scraped her knee and I offered to kiss her boo-boo, she said, not unkindly, "That won't do anything. Can I have some ice?" Later, I tried to convince her to go to bed on the night before vacation by telling her that the sooner she went to sleep, the sooner it would be morning, when we would leave for the beach. "Whether or not I go to sleep is not going to affect how fast time goes," she replied. My efforts to console her as she cried over a broken teacup from her tea set also were futile, but then her brother cheerfully informed her that "actually, *everything* is going to break. Even the whole universe is going to end!" She nodded at this and began to dry her tears. "And the sun will burn out," she said.

I never planned to write about private scenes like this; I kept my job as a science reporter separate from my life at home. But about a decade ago, a large spider spun a web in the corner of our kitchen window frame, and I grew to care for it. My friend and mentor, the science writer Ann Finkbeiner, encouraged me to write about this spider for a blog run by science journalists called *The Last Word on Nothing*. The blog's name comes from a quote by Victor Hugo: "Science says the first word on everything, and the last word on nothing." A number of professional science writers who want to break out of their routine use this blog as a creative sandbox (the blog's unofficial motto: "we write about whatever the &*%#$ we want"). I hadn't written a personal essay before. It felt risky. A reporter traditionally remains somewhat anonymous, and I had been a reporter for more than half of my life. I wasn't sure I *wanted* people to know what I thought about in my kitchen while I watched a spider. But once I wrote that piece, I began to write about other personal experiences, too—including ones that were more fraught.

Now, I'm not a scientist, any more than my children are. But it turns out that, like them, and for not dissimilar reasons, I felt compelled to experiment—to *essay*. While the stories I share in this book mostly deal with common occurrences that many people experience while growing up or parenting—like getting hit on by an older guy during adolescence, considering the possibility of abortion, calming a frightened child—I either experienced them or remember them as someone who has spent decades contemplating the history, culture, and values of the scientific enterprise. Despite what's taught in school about the scientific method, much of scientific inquiry, like poetry, involves play and metaphor and idiosyncratic obsessions and just plain fiddling around with mysterious things, things that are—to borrow a phrase from Walt Whitman—transient and strange.

Part I

The symbol of a tornado

I'm lying in bed between my young son and daughter, pretending to relax. We've just finished reading *Winnie the Pooh* and I've turned off the light. My son's body is disturbingly still. "Nell," he says in a small voice, "I'm thinking about tornadoes."

A few weeks earlier, he'd asked me what a "twister" was, after listening to a recording of a children's novel called *Twister on Tuesday*. I was excited that I—lucky me!—got to introduce him to this awesome natural wonder. I wanted to show him a spinning column of clouds snaking down to the ground because, I thought, it's the kind of phenomenon that inspires both philosophers and physicists. In retrospect, I'm amazed—although maybe I shouldn't be—that my mind apparently blanked on tornadoes' most important, perhaps even defining trait. I acted as though he'd asked about a rainbow. I searched the Internet and found a three-minute-long National Geographic video called *Tornadoes 101*. My children watched it, while I watched their eyes.

"Tornadoes," the narrator intoned, "big, funnel-shaped clouds that can rip through a community and leave a wake of destruction. They can form in seconds, change direction in a heartbeat, and their devastation can last a lifetime."

I saw their eyes widen almost imperceptibly and realized my mistake. "The vortex extends towards the ground. As it picks up speed, anything in its path can be ripped apart or thrown into the air," the narrator continued, over shaky footage of wrecked homes and flying debris.

My children wanted to watch this video again. And again. And again. Now, like a team of tiny storm chasers, they can rattle off a truly impressive number of tornado-related facts. And every night, just before sleep, my son's thoughts turn to twisters.

"Look," I tell him, trying to sound reasonable, "I don't think you have to worry about this. I've gone my entire life, forty-two years, without seeing a single tornado. I don't even know if Washington, DC, has even *had* a tornado."

"Tornadoes have been recorded in all fifty states," my son says, quoting the movie verbatim.

"Washington, DC," I retort, "is not a state."

"Yes, it is," he says, undaunted. It turns out that he was, in fact, correct, at least about tornadoes blowing through our city. I find a photo of one near the Washington Monument, looking like the white obelisk's upside-down cloud twin.

"Look," I say again, "I really think you can rest easy. It

was a beautiful day. I didn't hear anything about any kind of extreme storm in the weather report."

He's burrowed under the blanket, and says, sadly, "But what if the weather report is wrong?"

I concede that weather prediction isn't an exact science. But even though forecasters can be wrong when they try to predict things days in advance, I tell him, they are rarely wrong about how the weather will evolve over the next few hours. It's not as if they predict a blue sky and then, two hours later, a tornado rolls in! Even as I tell him all this, I feel a wispy cloud of doubt at the edge of my conscious horizon. Either reassured or just exhausted, he stops talking. Once he's asleep, I sneak away to go read on the couch downstairs.

A blue flash makes me look up just as my husband's cell phone starts buzzing with a Severe Thunderstorm Alert from the National Weather Service. Soon trees are bending in the wind, lightning crackles, and something starts pelting the windows. I think it must be rain, but the sound isn't right, and I realize it's hailstones. I vaguely recall that hail is associated with tornadoes, and I wonder helplessly how, if we needed to run for safety, I would even know. (I don't think to look up the Hailstorm FAQ produced by the National Severe Storms Laboratory until the next day: "Since large hail often appears near the area within a thunderstorm where tornadoes are most likely to form, you should assume a tornado could be nearby and seek appropriate shelter.") My husband goes out our back door, collects a bowl of hailstones the size of grapes, and puts it in the freezer. He wants to be able to show them to the children.

I don't wake my kids to go hide in the basement. And the

next morning, the sky is clear again. But the *Washington Post* is showing photos of trees ripped up around the Capitol, and I see this headline: "Super intense surprise storms slammed DC." The article says that "seemingly out of the blue, a lonely summer thunderstorm blew up" and blasted the city with "one of its most powerful thunderstorm events in years." And, it noted, "what made the storm particularly remarkable was that it was not forecast by the National Weather Service or any media outlet."

I think of the hail waiting in the freezer, like an icy tattletale. I ask my husband if he thinks it's a good idea to show this to our son. "Oh, I already told him," my husband replies. "He was very excited!"

....

My son picks up a flat package that arrived in the mail. He can tell it's a book and wants me to open it. "I don't think it's for you," I say warily. I open the package and hand him the thin blue volume. "It's a history book," I hedge.

He flips through it—he can't read—and immediately finds an illustration showing a twisting cloud beneath a dark sky. "It's a tornado," he says with surprise, and continues flipping through the pages as I admit, lamely, "Yes, it's a history book about tornadoes."

He finds a drawing of devastation in Illinois in 1883—it's a field of broken boards, with a locomotive in the distance—and another showing crumbled buildings from 1882. Then he reaches the end of the book and looks at the final page number, which says 90. "There's really more like 93 or 94 pages, because the pages with the pictures don't have numbers," he remarks. He hands me the book and walks away.

Tornadoes: What They Are and How to Escape Them was published in 1888 by the pioneer of scientific tornado studies, John Finley. He was a meteorologist with the Army Signal Service who gathered accounts of some six hundred tornadoes so that he could compile basic information, such as when and where they occurred. "The mystery of the tornado cloud has been swept away as the result of prolonged and thorough investigation," he wrote, with admirable can-do spirit. I find myself warming to him. Finley notes that "the tornadic disturbance is as old as the world itself, if we are to believe that the appearance of the atmosphere was coincident with the creation of the Earth," and cites descriptions of tornadoes that go back centuries. In 1762, for example, near Charleston, South Carolina, one tornado arrived "resembling a large column of smoke and vapor rushing over the earth with prodigious velocity, destroying everything in its path." Going back even further, God spoke to Job out of the whirlwind.

Finley was the first meteorologist to have the audacity to take these storms on as his life's work, living as he did in the nation that has the highest frequency of tornadoes—although, it should be said, the United Kingdom has more tornadoes relative to its land mass than any other country. His writings are chock-full of facts and bold assertions (everything from the etymology of the word *tornado*, which comes from the Latin *tornare*, "to turn," to the necessity for tornado insurance). Even though he handed out scientifically dubious advice ("under no circumstances, whether in a building or a cellar, ever take a position in *a northeast room, in a northeast corner, in an east room or against an east wall*"), he also strongly urged people to build "tornado caves," saying that shelter underground was the only place safe from a tornado's fury. He pointed out that his proposal had been ridiculed by people who found the idea of cowering in a hole undignified. "There is nothing to prevent any man from attempting the construction of a tornado-proof building or cage, something that will be aboveground and possess both strength and architectural beauty," he wrote wryly, "but I venture to assert that the man who thus essays to grapple with the tornado on its 'own ground' will not be one of the genus *homo* who has actually experienced a genuine 'twister.' "

On an experimental basis, Finley began issuing tornado predictions in 1884. But by 1887, the word *tornado* was banned from weather forecasts. The Report of the Chief Signal Officer stated that "it is believed that the harm done by such a prediction would eventually be greater than that which results from the tornado itself" because of the panic that would ensue. The Weather Bureau Stations Regulations of 1905 stated, "Forecasts of tornadoes are prohibited." The restriction remained in place until 1938, according to historian Marlene Bradford, who wrote that forecasters employed euphemisms such as "severe local storms" or, rarely, "conditions are favorable for destructive local storms."

I have no doubt what Finley would have thought of this. When I reach page 90 of his short but passionate volume—the final page number that, as my son pointed out, is not technically accurate—I read the words he wrote long ago: "Nothing is gained by trying to conceal the truth. Such a course begets indifference and negligence, which must eventually result in great evil."

My son is going upstairs to bed with his father, and turns to me to say, "I am worried about tornadoes." I glance out the window and reply, "I think you're okay for tonight, the weather is looking good." He pauses and says, stammering a bit, "Nell . . . what is the symbol of a tornado?"

The symbol of a tornado. What could he possibly be talking about? Then I understand and say, "You mean like on the weather report, what kind of picture do they use to show that there might be tornadoes?" "Right," he says, "I know they have a sun, and clouds with lightning, and snowflakes, but what do they use for a tornado?" I tell him I have never seen a forecast for a tornado so I have no idea if meteorologists have a symbol for tornadoes, but if they did, then I imagine it would look like a little corkscrew. "Did I mention I have never, in my entire life, seen a tornado?" I say. Satisfied, he goes upstairs.

Secretly, though, I have been looking at photographs of tornadoes. And even though I would have automatically and unthinkingly described them as funnel clouds, storm chasers use much more descriptive terms like cones, wedges, elephant trunks, needles, drill bits, stovepipes, and hourglasses. Photographs of tornadoes can be beautiful, but they've also historically been essential for scientists, because the way that tornadoes touch down and disappear makes them a difficult phenomenon to study. The first good estimate of the wind speed inside a tornado, for example, came in the 1950s, when a researcher tracked the movement of debris, frame by frame, in a movie that someone filmed of a tornado that hit Dallas, Texas. By looking at the distance traveled by the debris over time, he was able to estimate that wind speeds must have been as high as 170 miles per hour.

What is believed to be the first photograph of a tornado can be found in the archives of the Kansas State Historical Society. Described as "a long rope of purplish colored cloud," this twister touched down west of the town of Garnett in 1884 and was slow moving enough for a local fruit farmer and amateur photographer to set up his old-fashioned equipment. I study the photo. There's an unpaved street, wooden houses, and, in the distance, a trail of what would almost look like smoke going to the sky. But the trail looks smeared, apparently "touched up" by the photographer. Photographers back then commonly "enhanced" their photos in this way, but I can't fathom why, having obtained the first precious photographic evidence of a tornado, anyone would mess with it. Did the cloud look disappointingly insubstantial, somehow insufficiently terrible? The destructive power of a tornado, I have learned, has nothing to do with its appearance.

And what strikes me is that even though we all have a mental image of a tornado, and even though we now have all these photos and videos, even the experts don't always understand what they're looking at. In his memoir *Tornado Alley: Monster Storms of the Great Plains*, scientist Howard Bluestein describes his early experiences with storm chasing in the late 1970s. He was wild to see his first tornado. One hazy late afternoon, his team was driving down a paved road when "the silhouette of a huge cylinder" crossed the road in front of them. "It didn't look like any of the photographs of tornadoes I had ever seen, most of which looked like an elephant's trunk," Bluestein wrote. "Was this a real tornado, my first?" The cloud cylinder was rotating, but he thought it might just be a "wall cloud," a layer of cloud that extends down from a cumulonimbus but doesn't touch the ground, unlike the tornadoes it can spawn. After this mysterious apparition passed, he and his colleagues checked its path. They found downed power lines, trees with torn limbs, and a house missing a roof. Finally, Bluestein was convinced he'd seen a tornado—but the sight of the damage left him "queasy."

One of the most notorious tornadoes ever was the El Reno, Oklahoma, tornado of May 31, 2013. (Tornadoes invariably take the identity of the place they ravaged, and the date.) It was a violent, multi-vortex maelstrom more than 2½ miles across, with rotating winds of 300 miles per hour. This huge, unusually erratic tornado made sudden turns, expanded rapidly within seconds, and kept changing speed, at one point racing across the landscape at 55 miles per hour. It killed three well-known tornado scientists—the first deaths from professional storm chasing—and inspired much soul-searching among storm watchers, who have compiled videos of this monster taken from all directions. Some of the footage shows a swirling mass of darkness that fills the entire sky. I'd be hard-pressed to identify any part of this churning storm as the "tornado."

And I'm not the only one. Veteran storm chaser Skip Talbot, who wrote a long account of the El Reno tornado on his website, described how he waited for flashes of lightning that let him glimpse the funnel cloud. At one point he thought it was safely a couple of miles away. "I had no idea of the scale of what I was looking at, or the immediate danger that it represented," he wrote. Just then his partner noticed something odd about the rain immediately in front of them. "I paused a moment to watch it. The bands were careening at high speed from left to right. The huge column of rain shrouding what I thought was the tornado was rotating, and it was rotating at the speed of a tornado. The realization sunk in that the entire column of rain shrouding the funnel WAS THE TORNADO." They scrambled into their car and sped off, with the spinning storm following close behind, and at times gaining on them.

In the end, the El Reno tornado just disappeared. They all do. A tornado lasts from mere seconds to a half hour or more. Most last less than ten minutes. Then, for reasons scientists don't understand, it transforms back into ordinary clouds and drifts away. In its final minutes, it can look like a thin rope as it "ropes out." But a rope tornado can still be powerful—the winds can speed up as the funnel narrows. I look at one image, a white line of cloud stretching down, and think of the "magical, sometimes horrible" whale-line in *Moby-Dick*. This rope, attached to the harpoon, coiled in loops around oarsmen in their small boat. And when a harpooned whale bolted, this line would slice through the air like lightning, sometimes catching a man and dragging him overboard to his doom. Yet still the men would cheerfully row out to a whale, merrily bantering, seemingly oblivious to the "hempen intricacies" that encircled them like a hangman's noose. "But why say more?" wrote Herman Melville. "All men live enveloped in whale-lines. All are born with halters round their necks; but it is only when caught in the swift, sudden turn of death, that mortals realize the silent, subtle, ever-present perils of life."

When we go to the library, my son gathers all the books on tornadoes and piles them up on a table. He turns the pages, solemnly studying before-and-after photos. Here's a cute little cottage. Now here it is transformed into an expanse of broken planks and bent metal. His sister looks over his shoulder. "There's a teddy bear," she says brightly, pointing to a bit of pink.

At his request, I read the books aloud, adopting a flat, just-the-facts tone of voice. I read the words of a seventeen-year-old girl whose house tore apart above her while she cowered in the basement. I read how she gathered a few valuables before seeking shelter ("That's the last time I ever saw my room") and how the 200-mile-per-hour winds sounded like a jet engine screaming. "The fragments of red fabric are pieces of Megan and Matt's parents' bedroom rug," I read, deadpan, as my son's eyes dart across the photo, searching for red, and I wonder to myself what kind of parenting this is, anyway. My approach to this whole situation seems deeply flawed. " 'Tornadoes come in all colors, because they take on the color of materials they pick up,' " I go on as my son listens, rapt, to information that has got to be totally useless, at best providing a false sense of security and at worst just adding to his nightmares.

But who am I to judge? I've been doing my own reading. How does it help me to know that tornadoes have formed over snow-covered ground or that in 1842 some "scientist" did experiments that involved shooting chickens out of cannons, to verify reports of tornadic winds plucking the feathers off birds? Why am I poring over storm chasing advice that explains how to tell if you're in a tornado's path? ("If the tornado is not evidently moving to your right or left, but is simply getting larger with time, *then you're in the path*. This is not a good place to be.") So, because my son has asked me to, I keep reading tornado books to him, even though it feels sadistic. "A violently spinning, elephant's trunk-shaped tornado bears down on Interstate 70 near Quinter, Kansas," I say, reading even the photo captions without comment as he listens. "Less than a minute after this photo was taken, the twister crossed the interstate, tossing a car more than 300 feet, seriously injuring the driver."

That night, at a friend's dinner party, the kids go off to play, and I ask if anyone at the table has ever seen a tornado. No one has. "Well, I sort of almost did," says one guy I don't know, a heavyset man with a mustache who I believe is a physician. "When I was about ten years old my family went out to eat at a restaurant, and about an hour after we left a tornado came through and destroyed it." I stare at him. "It was completely obliterated," he says, carelessly.

I can reassure my children that tornadoes don't hit cities, he tells me, because cities generate heat that pushes them away. "But you probably already knew that," he says, in what feels to me like a patronizing manner. I don't tell him what I know, which is that hundreds of cities have had tornadoes—even London—and the only reason they're not hit more often is that suburbs and the countryside offer more area to strike. In 1953, for example, when future storm chaser Howard Bluestein was a young child playing outside at home near Boston, his mother told him he had to come in because she'd seen TV reports of a tornado in the area. ("As further inducement to getting me in the house," he wrote, "she told me that tornadoes snatch children up into the air and abduct them.") That tornado tore through Worcester, the second largest city in Massachusetts, killing ninety people, injuring around twelve hundred, and leaving ten thousand homeless. I can see that, already, the doctor's attention has wandered from tornadoes. He is staring at his iPhone, looking up the number of calories in a slice of pickle.

Many people in Worcester did not notice the 1953 tornado until it was upon them. They thought they were just having an unusually strong thunderstorm—and why wouldn't they? Even fifteen minutes before the tornado, when enormous hailstones fell and people marveled at their size, "Worcester residents did not know what they meant, and no one is reported to have taken them as tornado warning signals," according to an analysis written by an anthropologist named Anthony Wallace. It was commissioned by the National Academy of Sciences and entitled *Tornado in Worcester: An Exploratory Study of Individual and Community Behavior in an Extreme Situation*. Interviews with survivors revealed that only twenty-two out of fifty saw the funnel cloud, and only fourteen recognized it as a tornado. The others either didn't know what it was, thought it was smoke from a fire set by lightning, or considered the possibility of it being a tornado but dismissed the thought because the idea of a massive tornado hitting an East Coast city seemed ludicrous. Included in the report is a blurry, black-and-white photograph of the tornado taken by a news photographer when it was a couple of miles from the city. The image shows a tall, gray wall of turbulent cloud stretching up from the tree-lined horizon, with its top and one side cut off by the edges of the photo's frame. That's what a fourteen-year-old newspaper delivery boy saw before he ran into a nearby house to warn people that a tornado was coming. "Despite their skepticism," the report notes, "everyone took shelter in the cellar; the house was destroyed."

One night, back when I was dating my soon-to-be husband, we were driving to visit his parents, taking rural roads across Pennsylvania and watching lightning streak over farmers' fields. From the radio, I heard a sound that I'd heard all my life—the screech of the emergency broadcast system. I waited for the announcer to say this was a test, only a test, but, shockingly, he didn't. Instead, he reported a tornado warning for the very town where we were headed. Soon we arrived at the house where his parents' friends were hosting a party. It was summer and they'd been showing a movie in the backyard; everyone had been drinking wine and the mood was slightly boisterous. When we burst in, the sight of my twenty-something-year-old self, all panicked about a possible tornado, seemed to strike these retired academics as sweetly hilarious. I went down to the basement, but no one else did. I can't remember when or how I decided to come out.

I figure I might as well call up Howard Bluestein to ask his advice on my kid's tornado phobia. He's spent decades at the University of Oklahoma, and his efforts to put scientific instruments in the path of an oncoming tornado inspired the hit movie *Twister* in 1996, which in turn inspired a whole slew of young people to study severe storms.

When Bluestein was first getting started in his career, scientists had already made some progress in understanding tornadoes, helped by the invention of weather radar. Radar engineers who were trying to track enemy airplanes in World War II noted that their radio signals also bounced off raindrops and cloud droplets. What was annoying interference to them was of huge interest to meteorologists, who took the technology and ran with it. Before too long, they'd used radar and other measurements to make the first successful tornado forecast—although that seems to have only occurred because of a freak event. On March 20, 1948, a tornado unexpectedly hit Tinker Air Force Base in Oklahoma, flipping over airplanes and causing widespread damage. A pair of air force weather officers spent the next few days reviewing the weather patterns right before that incident, and as they did so, they realized that eerily similar weather conditions were developing once again. So, on the afternoon of March 25, with encouragement from their commanding officer, they took the unusual step of issuing a warning that a tornado was likely to strike the base. Just hours later, a tornado did. As unlikely as it was that two tornadoes would strike the same spot in less than a week, this success received widespread attention and made tornado forecasting seem like more than some pipe dream.

Another breakthrough occurred in 1953, when a radar engineer in Illinois noticed a weird hooklike pattern extending down from a thunderstorm image on his radar screen, looking almost like a curled and beckoning finger. He captured this strange radar signature on film and soon learned that this storm produced a tornado that seemed to be associated with the hook. That same year, a research radar station set up at the Massachusetts Institute of Technology captured images of the storm that sent a tornado barreling toward Worcester, scaring Bluestein's mother—and it also showed this odd curved shape. The so-called "hook echo" is now a classic sign that conditions could be right for a tornado. In 1964, the federal government set up the National Severe Storms Laboratory in Norman, Oklahoma, and began experimenting with the brand new "Doppler radar," a more advanced technology that could detect not only the presence of particles like raindrops but also their movement, revealing the wind patterns inside storms. These days, TV weather forecasters casually refer to their "Doppler radar" so often and so rapidly that it almost goes unnoticed in their chatter about whether to wear a raincoat. It's hard to imagine, now, what an astonishing new tool Doppler radar was for meteorologists a half century ago.

But they had to figure out how to use it. And that meant going out to eyeball storms and gather some ground truth that could be compared with the signals being received back at the Doppler radar stations. Although a few intrepid storm chasers had gone tilting at tornadoes in the 1950s and 1960s, those were just enthusiastic amateurs. Scientists like Bluestein were among the first to systematically hunt these monsters. Since 1978, when he first started chasing, he's witnessed hundreds of tornadoes.

During one memorable Oklahoma storm, in 1991, he and his buddies initially had trouble locating the tornado. "We could hear the tornado sirens in Billings wailing, and it was ominously dark to the south. Hailstones as large as baseballs bombarded the van. I didn't like what I saw. If we went south, we might drive straight into a tornado," he wrote in *Tornado Alley*. His team retreated and drove east. "We crossed the interstate and the tornado became more clearly visible. It was huge, and heading in our direction." His team found a safe place to set up a portable Doppler radar along a highway, and made close-up measurements as the tornado churned by just north of their position. They recorded peak winds of over 270 miles per hour—which was, at the time, the highest winds ever measured in a tornado.

When I get Bluestein on the phone, he tells me that my son has nothing to worry about. "It's very, very, very difficult to get near a tornado. Extremely difficult," Bluestein says. "We struggle just to even get as close as we do. So the chances are, if you're sitting in one location, to have one actually come and hit you is minuscule."

He says that over his lifetime, the science of tornadoes has advanced tremendously. "With our new rapid scan radars, we can scan an entire storm in five or six seconds, which for all intents and purposes is simultaneous," says Bluestein. "We can see where the tornado forms within this storm and how long it takes after the appearance of the incipient vortex to actually become a tornado at the ground."

Radar, however, can't measure temperature and pressure—key ingredients in any storm. What's more, experiments and computer modeling suggest that surface friction is important for the formation of tornadoes, but radar isn't good at revealing what's happening way down at the ground. "You can get down to maybe 50 meters, maybe even less than that sometimes," says Bluestein, "but the problem is that the radar beam gets contaminated by ground clutter."

So even though scientists can predict the formation of a thunderstorm with a persistent rotating updraft, a so-called "supercell," they aren't able to predict if it's actually going to produce a tornado and when that would happen. "Can we tell, if there are three supercells out there, which one is the one that's most likely to produce a tornado? Right now we just don't know," says Bluestein. And then there are the tornadoes that appear out of thunderstorms that aren't supercells. "Those are virtually impossible to forecast," he says.

I ask Bluestein what I should tell my son about his fears of total, random obliteration. "I'm kind of at a loss," I say. "What am I supposed to say, 'This isn't going to happen?'"

"You tell him what I tell my wife who worries," replies Bluestein. "There's so many things that can kill you. The chances of a tornado hitting, where you are, are ridiculously small. You should worry about getting hit by a car."

"Um," I say, imagining a new car-related terror, "I don't want my kid to start worrying about the car."

"I know, but you could say that the chances of you getting hit are very, very, very, very tiny," says Bluestein. "It's probably not going to happen. It's probably more probable than getting hit by a meteor, but you're a lot more likely to get hurt in an auto accident." He says this is what he tells not just his wife, but also a young nephew who fears tornadoes.

"And do you find," I ask, "that this is a reassuring line of argument?"

"No," says Bluestein, and we both laugh.

I've watched videos of Bluestein talking about tornadoes, and in one he mentions a book he had when he was a kid. Simply called *Weather*, it was published in 1957. After our phone call, I order a copy from a used bookstore. The illustrations on the cover include a dramatic tornado cloud in black and white, and the table of contents has a chapter called "The Atmosphere: Restless Ocean of Air." I think again of *Moby-Dick*, with its three chapters devoted to "The Chase" of the white whale.

I flip through this vintage nature guide, imagining Bluestein gazing at these colorful illustrations when he was the age of my son, before storm chasing was ever something he imagined. Despite the cover picture, this book has almost no information on tornadoes. And some of its paltry facts are flat out wrong, like when it states that "an average of 124 tornadoes hit the United States each year." (With improved detection, researchers now know that the real number is more like a thousand.) One page features illustrations showing the development of a tornado, starting with "mammatocumulus clouds form." These rows of softly rounded clouds, named after breasts, are associated with severe weather but have nothing to do with tornadoes, as I learned while reading the book that Bluestein wrote as an adult, which dismisses the connection as "folklore." I open *Tornado Alley* again and look at the photos he took of mammatus clouds. Next I gaze at a "shelf cloud" and a "roll cloud," and then, a few pages after that, under another cloud photo, he has written: "The 'whale's mouth' as seen behind the gust front of a thunderstorm." A "whale's mouth" turns out to be a kind of scalloped cloud formation found in dangerous storms. To me, it does not look particularly whale-like, but it also is a kind of cloud I've never seen. I keep looking through Bluestein's book, reading about his chases and studying his photos, trying to see what he saw. Toward the end, in a section about avoiding getting hurt, I come across this sentence: "The most difficult situation is the approach of a tornado late at night, when most people are sleeping."

It's bedtime and my son is once again filled with fear that a tornado will come and destroy our home. I say that if a tornado came to our city, as unlikely as that would be, it would probably be the kind of weak storm that would only break some tree limbs and maybe damage the roof. He is quiet for a moment, and then says, "You mean like an EF1?"

He's referring, of course, to the Enhanced Fujita Tornado Damage Scale, which rates the strength of a tornado based on how much damage it caused. Every kid's book on tornadoes has some simple illustration showing the basic idea behind this scale. The enhanced version is more complex, with ways to assess damage to twenty-eight different kinds of buildings and trees. So, for example, an assessor could look at an "automobile showroom" and determine which of eight potential degrees of damage it had suffered, from just "broken glass in windows or doors" to "complete destruction of all or a large section of the building." The trouble is (and this is never discussed in the children's books), a tornado can sweep across a barren area and leave behind little visible destruction. The killer El Reno tornado was initially rated an EF5 because wind speeds of over 200 miles per hour had been recorded by Doppler radar. But then, based on the damage survey, the National Weather Service controversially downgraded it to a mere EF3. The move seemed almost disrespectful to this powerful tornado that swallowed up three experienced storm chasers, and critics pointed out that the Fujita Scale was supposed to be a proxy for when winds couldn't be determined.

But meteorologists have used the Fujita Scale to rank tornadoes ever since a storm researcher named Ted Fujita came up with this approach back in 1971. He knew it wasn't perfect, but it was the best he could do given that wind speeds above 150 miles per hour destroyed his measuring devices. Fujita, known to his friends and the public as "Mr. Tornado," was famous for taking aerial photographs of wreckage and then using those photos to make uncannily perceptive deductions about how a storm's winds must have blown. He's been called "one of the greatest meteorological detectives of all time." One colleague wrote about going with Fujita to Grand Island, Nebraska, in 1980, after a tornado-spawning thunderstorm had pummeled the city for more than two hours. "This storm produced the most complex damage patterns imaginable," the researcher wrote. "No one but Ted Fujita could have sorted them out." Fujita reconstructed the paths of seven distinct tornadoes, three of which were rare anticyclonic tornadoes, meaning they spun in a clockwise direction rather than the counterclockwise direction that's usual in the Northern Hemisphere.

Much of Fujita's autobiography, a quirky book called *Memoirs of an Effort to Unlock the Mystery of Severe Storms*, is devoted to page after page of photographed devastation, plus maps he created to show storm tracks. One map is covered with loops that represent his estimated trajectories of six 24-foot, 700-pound steel I-beams that a tornado picked up from an elementary school in Bossier City, Louisiana, in 1978 and flung through the air like missiles. One of the I-beams ended up in the backyard of a house, sticking out of the ground at a 23-degree angle. There's a photo of Fujita hanging off the end of this steel beam, using his own weight to test how firmly the tornado had driven it into the dirt. "It did not move even a millimeter," he reported. He recalls how, in his childhood days in Japan, he was taught that "four fearful things are, in the order of fear, Zishin, Kaminari, Kaji, Oyaji (Earthquake, Lightning, Fire, and Father)." When he came to the American Midwest, he realized that the primary fear there was not earthquakes but tornadoes.

Reading his memoir, though, I notice that his life was shaped by another fear. His father died in 1939, and his final words to his son were an order to attend Meiji College even if he was admitted to a college in Hiroshima. "His inspired final instruction may have saved my life because, had I attended the Hiroshima College, I could have been in Hiroshima when the first atom bomb exploded over the city on August 6, 1945," Fujita wrote. Instead, as his father had wished, he attended Meiji College and was hired there after graduation to teach physics. After the second atomic bomb dropped on Nagasaki, he learned that "the initial target of the second bomb was the Kokura Arsenal, 4.3 km (2.7 mi) east-southeast of our college. I clearly remember hearing a series of air-raid sirens on the day of the bombing, but the aircraft was not visible due to a thick layer of stratus clouds. Evidently the bomber flew away to Nagasaki while we were hiding in a bomb shelter next to the physics building."

A month after the nuclear explosions, Fujita visited both destroyed cities and saw the effects of the shock waves on trees and buildings. His keen eyes noticed bamboo and pine trees flattened and broken off in horizontal directions away from Ground Zero in a giant starburst pattern. "After coming to the United States," he wrote, "I photographed from low-flying Cessnas a large number of damage areas in the wake of tornadoes. Unexpectedly, I came across these starburst patterns of uprooted trees." A tornado typically left a swirling pattern. He realized that the unusual damage must have come not from a twister but from something that was its exact opposite—a "downburst." A downburst, or microburst, is a powerful downdraft that, instead of sucking things up, pushes things violently down. Critics scoffed at the idea, but Fujita and colleagues gathered evidence proving that he had discovered a previously unrecognized meteorological phenomenon, one that explained some mysterious and deadly crashes of airplanes during thunderstorms.

I'm reading about Fujita while on vacation at a beach house, and one night my kids find a book by Dr. Seuss called *The Butter Battle Book*. They bring it to me to read as a bedtime story. I am astonished to learn that it's a tale of an arms race between two warring communities, the Zooks and the Yooks, who develop increasingly elaborate weapons. At the end, both sides possess a doomsday device called the Bitsy Big-Boy Bomberoo. It's unclear if they'll drop these bombs—just in case, the Yooks have gone into hiding underground—or just exist in the suspended animation of mutually assured destruction. My children giggle at this silly story. I feel a sinking sense of dread, knowing that at some point they'll have to learn that the Bitsy Big-Boy Bomberoo isn't just some outlandish Dr. Seuss invention. I grimly wonder if I'm the one who will have to do the explaining. I glance at the publishing date—1984—and realize that this book came out when I was ten years old, about the same time that my brother won a decorated mirror at a dinky carnival. This mirror showed the Olympic rings, with the red one off by itself, and beneath this were the words: "Let the

Russians play with themselves." Back then, I did not understand the double entendre, but I knew that we were supposed to hate and fear the Russians. The threat of nuclear attack was always there, almost unnoticed, like air. Even now, the missiles of my childhood lurk in hidden silos, preprogrammed with their intended trajectories, one of them surely aimed straight for the nation's capital, the city that my children call home. I think about the people living in Worcester in 1953, who never thought of tornadoes but who were preparing for nuclear Armageddon, with "duck and cover" lessons taught in the local schools. The main reason that Worcester's tornado was studied so closely was to learn how people would react after an "extreme situation," like a sudden blinding flash followed by a mushroom cloud.

....

My daughter is sitting on my lap and eating a peanut butter and honey sandwich. She looks up at me with her big brown eyes, and says out of nowhere, "If a tornado comes we go to a storm cellar, right?"

I smooth her hair back from her face. She is three years old and has blond curls. Her brother had this kind of hair at her age, but as he got older the ringlets disappeared and now he has the same straight brown hair that I have. My daughter isn't wearing a shirt, and between her thin shoulder blades I see the small birthmark that I sometimes kiss. Now I touch it and tell her, lightly, "Yes. And if we don't have a storm cellar, which we don't, we go down to the basement."

"And if you don't have a basement you go to an interior

room!" my son pipes up from the floor, where he is building a spaceship out of Legos. "With no windows! Like a bathroom. Or a closet."

"When the sky turns green that means a tornado is coming," my daughter persists, gazing up at me. "Right, Nell? Right?"

* * *

In the kitchen of the house where I grew up, we had a phone attached to the wall, and the receiver had a long cord, a twisted stretchy thing that hung down in a tangle. In 1988, a few days before Valentine's Day, I was thirteen years old and making candies to bring to school for the holiday. The recipe involved melting chocolate chips in a pan, but I don't remember anything else about it. I was wearing a white terrycloth robe. My brother was in the basement. I could hear the wail of his electric guitar as he practiced rock songs over and over. The phone rang. Did it make me jump? I picked it up because I always did.

"Hello, may I speak to William Boyce?" a man's voice asked.

"He's not home," I said. "Can I take a message?"

"I'm calling from the hospital," the man said, "this is in regard to William Boyce, senior."

"You mean Gene?" I said, feeling a shot of panic, thinking, *My grandfather is in the hospital.*

Yes, the man replied smoothly, Gene.

The man on the phone wanted to know who was speaking. I gave him my name and my age. He wanted to know who else was home. I told him about my brother. "Call to him, but don't leave the phone," the man said. This request should have seemed odd, but I obeyed without question because I was raised to be polite and obedient to my elders. I screamed and screamed my brother's name, as loud as I could, but he couldn't hear me. The man asked where my father was, and I said he hadn't come home yet from work. He asked where my mother was, and I said she was at the town hall, for a committee meeting.

"No, she's not," he said. He sounded amused. "She's with me."

I was looking at the phone book in front of me on the counter. It was like some kind of drug had been injected into my veins. I went numb. He described how he waited for her in the parking lot and pushed her into his car. The phone book became an alien object that no longer made any sense.

Many people in Worcester did not understand until the final seconds that the storm bearing down on them was "different." That's when the shriek of the tornado could be heard, the air filled with mud, windows and trees began to break, and, as Wallace later put it in his report, "an almost intuitive awareness that disaster was upon them tripped off emergency action even without rational understanding of the event." People ran, ran for the basement, ran for a closet, ran for their children. Some even tried to run outside.

I didn't run. I stood there in the kitchen, in my bathrobe, with the long tangle of telephone cord wrapped around me, as the man ordered me to talk like we were having sex. "I can't do that," I stammered, trying to stall, "I don't know how, I've never had sex." This made him angry. He had my mother, he said. He told me what to say. I did what he wanted. Then he said he was coming for me, and hung up.

Did I drop the phone? Did it dangle by its cord? Did I hear

a dial tone? That's when I ran. I ran to the basement, to my brother, who came up and called the police. Since the police station was across the street from the town hall, an officer was dispatched to retrieve my mother from the meeting. When I heard her familiar voice over the phone, sounding confused and alarmed, I collapsed to the floor and sobbed. But right behind the relief came another huge feeling that I couldn't name, at least not then.

In Worcester, after the first impact from the tornadic winds, a wave of low pressure created an eerie "floating" phenomenon. One family had been cooking dinner, with potatoes just put in to bake, and the daughter recalled how "the potatoes came out of the oven and went over and hit my daddy on the head." Eggs floated out of crates. Furniture rose up and moved across the floor. Homes exploded, with roofs lifting off and the walls falling away: "Informant after informant describes how, after a few moments of raging wind, the whole building seemed to dissolve, in slow motion, and they would find themselves sitting in quietness and relatively uninjured in the yard, surrounded by pieces of house." This surreal calm only lasted twenty to thirty seconds, and then, as they sat bewildered and unprotected, the back wall of the tornado slammed them with another blast of debris and wind.

I didn't tell anyone exactly what had happened during that phone call. But after that, I couldn't be in the house alone without going cold and shaking. And I was alone in that house almost every day. I sat for hours, frozen with fear, listening to every slight sound. I refused to answer the phone; my family had to invent a special way of ringing once, then hanging up, then ringing again in a few minutes so that I would know who was calling. When the doorbell rang, I hid. One day at school I started crying and couldn't stop, so my mother was called to come get me. I don't think she had any idea what was going on or what to do. As she drove home, with me weeping beside her, she asked, in an exasperated tone, "Do we have to sell the house? Do you want us to sell the house and move?" I know now that the source of her frustration had to have been her own inability to help, but at the time, I just felt defective. Outside, I looked unchanged. Inside, everything had been violently seized up and then thrown back down in a heap. It felt as though there had been a test. There had been some kind of important test, and I had failed. I was staggered by the shame.

Almost three decades later, I struggled to explain this phone call and its aftermath to a psychiatrist. His office was small and dark, like an underground cave, and he was much older than I was. He sat across from me in an armchair, calm and patient. At one point I realized that he didn't seem to be aware that any part of the call had been sexual, although I'd thought I had made that clear weeks before. "No one understands. It was just a *phone call*," I told him, bitterly. "No one understands how words can change a person. Isn't that what you do, change people by talking to them? What do you think you could do if you took all your powers of empathy and insight and twisted them into something evil?" He wanted to know why this, and certain other things that had happened, did not leave me feeling angry. I explained that they seemed almost like natural disasters. The phone call was so weird and random that it couldn't be personal; if it hadn't been me, it would have been someone else. He listened attentively, slipping off his shoes, as he often did, and rubbing his feet together, in their

Gold Toe socks. "Nell, this idea of yours that it's not personal," he said, casually throwing one arm back over his head as he stretched slightly and shifted in his chair. "For lack of a better word, that's just wacky."

We have just read some *Frog and Toad* stories and now the light is off. The children are supposed to be trying to sleep, and I lie between them, tense, waiting for the madness to begin. "Nell," says my son, "I'm thinking about tornadoes. I can't stop thinking about tornadoes."

"Actually," I say, "You can. You *can* stop. That's because you can control your thoughts. You can force yourself to think about something else, something nice. Like you could remember being at the beach and what it felt like to ride the waves, and the hot sand under your feet, and the sun beating down on you."

He ignores this pathetic advice. "When a tornado comes, Roxana and Mihanna won't lose many toys," he says, referring to some friends who live a few doors down. I am perplexed, but then I remember that their playroom is in their basement. "I wish my bed was in the basement. We need to sleep in the basement," he wails, with mounting panic. I briefly consider our basement, infested with crickets and moldy rubbish that should be thrown away. I don't want to set a precedent of sleeping in the basement. "Don't I keep you safe? Would I ask you to sleep in a place that was not safe?" I ask, realizing, even as I say the words, that these questions are inane and inadequate, even insulting. (*Do we have to sell the house? Do you want us to sell the house and move?*) He doesn't answer. Instead, he gets up and goes to the window, pulling back the curtains so he can search the sky.

"I'm scared I'm going to die in a tornado," says my daughter, matter-of-factly. *Jesus Christ*, I think, *she's three years old*. But I reply evenly, "Oh honey, I really, really don't think that's going to happen." She says her leg hurts, and flinging it over my waist, she asks me to rub it. Both kids have started complaining of leg pain at night, and after doing some research I decided this fit under the vague diagnosis of "growing pains." (The Mayo Clinic's website unhelpfully notes, "Although these pains are called growing pains, there's no evidence that growth hurts. Growing pains may be linked to a lowered pain threshold or, in some cases, to psychological issues.") My daughter shoves a stuffed red monkey in my face and says, "Will you take care of monkey all night? Will you rub my leg all night?" I answer immediately and say, "Yes. Yes, I will. All night long, I will take care of your monkey and rub your leg," despite knowing I will creep away the moment they go to sleep. I'm not sure why I'm suddenly lying to my children.

My son comes back to bed and flings himself down. I think he's sleeping but then he cries out in fear, "When is the tornado coming?" How do you answer a question like that? The truth is, I don't know. But I feel compelled to say firmly, "It's *not*. It's not coming." He turns his back to me, faces the wall, and whimpers. I feel like things are beginning to spin out of control, like we're coming closer and closer to something we're not prepared to face. I try moving in a new direction. "Look, I don't think a tornado will come tonight. But if it does," I say soothingly, "we will be okay because we know what to do.

Your daddy's phone will buzz with a weather alert and he will come get us and we will all go down to the basement, where we'll be safe." These words feel cowardly and weak, even contemptible, and I hate myself.

My son does not appear to notice. "But Nell," he says, his voice trembling, "we will only have thirteen minutes of warning. That's not enough time to get all the things I need to bring down to the basement." I say nothing as I consider his memorized fact. I know that thirteen minutes of warning is just an average—we might have a lot more time or a lot less—but that's not necessarily reassuring information. Plus, I don't even know if he fully understands the concept of "average." He is, after all, only six years old.

He starts listing everything he will need to bring to safety—his stuffed rhino named Spike, his model of the Millennium Falcon, the special house he built for his Tiger, all his Lego spaceships and begins to quietly weep. I tell him that all this can be replaced, that I will replace it all. If the plastic Millennium Falcon gets blown to pieces by a tornado, I will buy him a brand new one.

"What about my allowance?" he says, thinking of the $64 he has saved in a plastic mayonnaise jar. "I won't have time to get my allowance."

I will replace all the money, I tell him.

"All at once," he asks, abruptly sounding calmer, "or week by week?"

All at once, I say, all the money, all at once. And we'll get a new jar to put it in. He snuggles closer and I stroke his hair. Speaking softly, like I'm singing a lullaby, I promise to buy a new jar of mayonnaise. I will empty the jar and put it in the dishwasher so it will slightly melt and get misshapen, just like the old one. I will write his name on the side with a red Sharpie marker, so it's just like the jar he has now. I'll put the crumpled dollars in, and then the coins, and it will look the same. It will look exactly the same. I will fix everything, replace everything, make everything perfect so that in the end no one will know—not even you, I silently tell him, and myself, for I have stopped speaking aloud. My body feels heavy on the mattress, my heart beats like a promise as something inside whispers *you will forget, you will begin to forget, you will begin to believe you can actually forget.*

His breathing grows deeper and steadier, and my daughter's restless flailing becomes less frequent. When all is still, I carefully extricate myself from the tangle of their arms and slip out of the bed. I straighten the white, fluffy comforter and tuck it around their bare legs, their poor legs that ache with the pain of growing. I place the red monkey between them, with his little head on a pillow, and tuck him in too. Years from now, when my son no longer trembles at the word "tornado," when he calmly goes to sleep alone after reading comic books instead of urgently checking the skies, I will remember this night. I will remember how I picked up their empty milk bottles, put their dirty clothes in the hamper, and took one last look around the darkened room. I saw the painting of wolves on the wall, the scale model of the USS *Constitution*, and books scattered across the rainbow-colored rug. There was nothing else I could do—*was there anything else I could do?*—so I carefully stepped out of their room and closed the door.

What else is there?

My father, from his hospital bed, asks me, "What are you wearing?" I look down at the blue sweatshirt I'd bought the day before. I'd gone to visit my parents with only an overnight bag, ostensibly to pay them a visit but also because I'd been worried about my dad's health. I was right to worry. He was so unsteady that he fell down right in front of me, next to our car in a neighbor's gravel driveway. I took him to the emergency room and he was hospitalized. Knowing I wouldn't be going home anytime soon, I went out and bought myself extra underwear and socks and this sweatshirt.

"I just got this at Target," I say. "I had to look through all the trendy clothes to find something I could plausibly wear without looking ridiculous." I try to be entertaining and cheerful during visiting hours, because it's an otherwise grim scene. My father now requires oxygen and can no longer swallow liquids.

But I have misunderstood; he doesn't care about my shirt. His voice sounds hoarse and weak. "What's . . . what's your decoration?" he says, as if searching for the word.

I reach up and touch the lump of gray metal hanging from a chain around my neck. I say, "it's a meteorite."

My father's eyes widen. His expression is solemn.

"I wanted a rock from outer space, so I bought one and made it into a necklace," I add.

My father says nothing. I'm not sure how much he's actually understanding.

"It also reminds me of Lou," I say, referring to my grandmother, his mother. "She always used to wear all those necklaces with big pendants." This makes him smile.

My father is eighty-three years old. The meteorite hanging from my neck fell to Earth when he was nine. At 10:38 a.m. local time on February 12, 1947, a fireball brighter than the sun appeared over the Sikhote-Alin Mountains of eastern Siberia, as an estimated 110 tons of mostly iron and nickel from the asteroid belt between Jupiter and Mars ripped through the atmosphere. Some eyewitnesses assumed it must be a nuclear explosion. One of them, an artist named Pyotr Medvedev, made a painting of the smoke trail it left across the sky, which looked like a long funnel cloud that ended in fire. Because members of the Russian Academy of Sciences hurried to investigate, tramping through the snowy taiga, this became one of the most extensively studied meteor falls in history. So much metal got littered across the mountains that pieces aren't hard to come by. The one I bought is dark gray and about 2 inches long, and it's covered with "thumbprint" marks, which are oval depressions created by hot gas as the metal plunged down.

I had a jeweler put a thick band of silver around the top with a loop, so I could wear it on a chain. The bright, smooth silver contrasts with the dark, lumpen meteorite. The heavy weight feels reassuringly *there*, and I wear it every day, often reaching up to touch it. Only later did I realize that I'd unknowingly made myself an echo of what is possibly the most famous rock on Earth: the Black Stone.

The Black Stone is enshrined in the wall of the Kaaba, the cube-shaped building at the Grand Mosque in Mecca, Saudi Arabia. There, in the eastern corner, an oval silver setting surrounds darkness that, from a distance, appears to be a single large, black stone. Pilgrims touch the stone if they can get close enough, or just point to it as they circle the Kaaba. The Black Stone's association with the Kaaba dates back to ancient times, before the birth of Islam, although Mohammad reportedly placed the stone in its current position after it had been removed for building repairs. One tradition holds that the Black Stone fell from heaven to Adam and Eve, that it was originally white, and that the sins of humanity turned it dark. Some say the angel Gabriel gave the stone to Abraham.

Wherever it came from, the Black Stone is no longer one stone; over the ages it's been broken into pieces. High-resolution photos released by the Saudi Arabian government in 2021 show that the silver setting actually contains a kind of matrix that looks to be reddish orange and resinous, like frankincense. Embedded within this matrix are seven fragments of dark gray rock. No one knows what kind of rock. The Black Stone has never undergone a modern geological analysis, although occasionally scientists have speculated about its origins. Many assume the Black Stone must be a meteorite, because the notion that it "fell from heaven" fits in with a long history of meteorites ending up in religious shrines. One geologist who looked at the close-up photos for me said it didn't necessarily look meteoric, but it's hard to tell from just a photo.

Regardless of what it is, "it is venerated by billions of Muslims," another geologist, Faroud El-Baz, told me. El-Baz is famous for his work with NASA during the Apollo program; he helped choose the lunar landing sites to maximize the scientific reward. He thinks the Black Stone fell from space, and he has a special connection to this rock. In 1974, El-Baz told me, he visited Saudi Arabia with his family and saw the Black Stone, with the pilgrims longing to touch it. This gave him an idea. He had recently left NASA, because the Apollo program was coming to an end, and had taken a new job at the not-yet-open-to-the-public National Air and Space Museum. He realized that this new museum ought to feature a "touchable" moon rock.

It took him a year and a half to convince skeptical NASA officials that one of the nation's priceless Apollo samples, which were otherwise kept in a secure vault, should be put on display for random grubby hands to grope. But El-Baz's faith in the emotional power of touching an outer space stone was vindicated when the museum opened in 1976. "It has been a most attractive exhibit to young kids," El-Baz told me. And adults love it too. Millions of people from all kinds of backgrounds have lined up for the chance to reverently run their fingers over this piece of dark gray, iron-rich, lunar basalt.

I'm one of them. I understand the allure. I even wear a space rock around my neck. But there in the hospital with my father, watching him sleep, clutching the meteorite in my hand, I know that almost no one would notice if the museum swapped out its moon rock for some nondescript gray one from the gutter outside in the street.

I wonder if any part of my father is aware that he owns a fragment of the Moon.

For Christmas, just a few weeks before, I gave him a piece of a lunar meteorite. I had been astonished, when purchasing my own meteorite, to learn that I could also buy fragments of the Moon that had made their way to Earth after some cosmic impact sent them spinning off into space. Scientists were surprised, too, back when they first realized that rocks from the Moon and Mars sometimes end up on our planet. The asteroid belt is the source of almost all of the tens of thousands of meteorite finds listed in the Meteoritical Society's official database. In 1981, however, researchers in Antarctica found a rock that looked suspiciously like rocks brought back by the Apollo astronauts, and a chemical analysis confirmed that their composition matched.

My whole life, I have struggled to find Christmas gifts that might genuinely please my father; my mother enjoys buying toys and gadgets and piling up gifts under the tree, but my father always claims to want nothing. I feel the same way. And yet I always wrap up some token gift for him, like a cardigan sweater or a pair of socks. A $150 cube cut from a lunar meteorite seemed like a plausible gift. While my father has never expressed much fascination with space, he did work for NASA in the 1960s, calculating trajectories for the Apollo program.

I put the cube of gray rock inside its little plastic specimen box, along with a descriptive card, inside a larger box to wrap it. In the chaos of Christmas morning, as the children ripped open dozens of gaily colored packages and threw the wrapping paper into the fireplace, my father opened his present and thanked me for the chocolate. "No, no," I said, "it's not chocolate, that's just the box I used to wrap it, the real gift is inside."

Then I got distracted by the children and my own gifts. I didn't see his reaction.

It had been an unusual, distressing holiday. Something was going on with my father's brain. He was having "spells" that seemed to be a kind of seizure. During Christmas Eve dinner, he stared into space and lost consciousness. We called 911. He recovered quickly, but the spells returned after a few days, and the hospital agreed to keep him for tests. When he left with a new medication and a walker, he seemed to be improving. But I worried. That's why I came back a month later to check on him, and saw him fall. It turned out he needed to be hospitalized again, this time with a more serious illness.

One afternoon, as I was getting ready for my daily pilgrimage to the hospital during visiting hours, my mother reached into a kitchen drawer, the one filled with pens and notepads and postage stamps. She pulled out the tiny plastic box, with the gray rock inside, and held it up for my inspection. "Nell," my mother said, baffled, "Do you have *any idea* what this is?"

"It's a meteorite," I told her. "From the Moon. It is literally a piece of the Moon. I gave it to Dad for Christmas."

"Oh," she said. She regarded it briefly. Then she put it back in the drawer, next to a box of envelopes and the good scissors.

It did not surprise me that my mother had been mystified by what looked to be a carefully preserved piece of concrete.

For a rock to be venerated as some unearthly treasure or message from the gods, there needs to be some clear sign that it's not like every other random hunk of matter. At the Smithsonian, curators hung a sign that says "MOON ROCK" over the touchable moon rock. And historically, "the regard in which meteorites have been held depends wholly on whether their fall was observed or not," according to folklorist Oliver Farrington in his 1900 article "The Worship and Folk-Lore of Meteorites." "It was always the fall and the phenomena attending it which impressed the observer, and not any peculiarity in the stone, if found alone."

When found alone, meteorites tended to be treated like any other rock, either ignored or exploited as raw material. In Kiowa County, Kansas, for example, farmers would periodically turn up chunks of a meteorite in their fields and put them to "all sorts of base uses," like holding down stable roofs and rain barrel covers, noted Farrington. In Staunton, Virginia, around 1858, one man tried to sell a piece of what turned out to be meteoric iron for a dollar, but "being unable to do this, he threw it into a backyard, where it remained until it was built into a stone wall. There a dentist discovered it, and found it very useful to hammer metals and crack nuts on." Inuit communities in Greenland long relied on giant iron meteorites for metal. They'd knock off pieces that could be hammered flat and used as arrowheads or knives.

This utility-focused attitude was matched by the indifference of the world's top scientists, who, for centuries, simply failed to recognize that some rocks on the ground might be extraterrestrial. The idea of rocks falling from the sky seemed ludicrous, and claims to the contrary got denounced as superstition or balderdash. "Eighteenth-century physical theory dictated that pieces of matter could not fall from interplanetary space," wrote John Burke in his definitive account of meteorite history, *Cosmic Debris: Meteorites in History.* Educated thinkers knew that space beyond the Earth contained only the great heavenly bodies and the ether, an invisible element that held planets and stars in their celestial spheres. With that accepted as an established fact since the days of Aristotle, many scholars assumed that any fireballs seen in the Earth's atmosphere must be gaseous "exhalations" from inside the planet that had escaped and got ignited. Other researchers suspected that the bright streaks could be "unusual manifestations of atmospheric electricity."

That's why President Thomas Jefferson pooh-poohed the claims of Yale scientists who, in 1807, collected what they said were newly arrived rocks after residents of Connecticut heard booms and saw a glowing in the sky. "It is easier to believe that two Yankee professors would lie than that stones would fall from heaven," Jefferson reportedly remarked. That quote may be apocryphal, but when one correspondent offered to send the president a fragment of this meteorite for analysis, Jefferson wrote back, "It may be very difficult to explain how the stone you possess came into the position in which it was found. But is it easier to explain how it got into the clouds from whence it is supposed to have fallen?"

Over in Europe, though, evidence was growing that rocks

fell from above, possibly even from outer space. That's because the tools of analytical chemistry had advanced enough for researchers to compare the elemental makeup of regular ol' rocks with stones associated with sky fire. In 1802, the English chemist Edward Charles Howard found that fireball-associated rocks had elements in common, such as nickel, which is rarely found on Earth. Then in 1803, just as the meaning of this new discovery was being debated in Europe, a huge meteor shower scattered thousands of stones over L'Aigle, France, not too far from Paris.

The French Ministry of the Interior sent a physicist named Jean-Baptiste Biot to investigate. His thorough report, written in an engaging, narrative style, included the testimony of eyewitnesses and mapped out the location of the fall, noting the broken tree branches and holes in the ground. What's more, Biot reported that all of the recovered fragments shared the same mineral makeup. Their composition was similar to stones associated with previous fireballs and did not match that of the rocks usually found around L'Aigle, where Biot found no nearby foundries or volcanoes that might have belched something out. His report provided "crushing evidence that stones fell from the sky," wrote Burke.

And so it is that these days, a geologist who gets shown a meteorite is very likely to recognize it for the out-of-the-ordinary specimen that it is. Most of the supposedly alien stones they see, however, come from right here on Earth. Eric Twelker, the prominent meteorite dealer who sold me my meteorite, as well as the one I gave to my father, once said in an interview that the biggest downside of his job was all "the people with driveway rocks who write every day asking me to buy their valuable 'meteorites.' I try to be polite and identify the rocks as best I can. I don't like to discourage them because once in a while a real one pops up."

He's especially skeptical of anyone who comes to him with a rock and claims to have witnessed its fiery descent. That's because all but the largest meteorites stop burning when they're miles up in the sky. Small rocks are impossible to spot from far away as they drop down through the clouds, going a couple hundred miles per hour. The eventual impact makes a dent or a hole rather than some awesome crater. So even though Hollywood fills the public's imagination with scenes of Earth-threatening fireballs raining death and destruction, most of the time the arrival of a space rock is a subtler experience. According to Twelker, those lucky enough to experience a fall will hear a "whoosh" and then a "thunk." And then a rock is there on the ground.

....

In the hospital, I start making a mental list of things that my father mentions wanting. He'd like some takeout Chinese beef with broccoli. He'd like to see my brother's new house that's being renovated. He'd like to take me to the town's library to see a plaque with my parents' names at the reference desk. He wants us to have dinner at a restaurant called Alfie's.

I think about this modest wish list. I don't know if my father will ever be able to enjoy beef with broccoli again. He's on a diet of pureed foods, which he struggles to eat.

I look around the hospital room. Next to the television is a whiteboard where the nurse has scrawled her name. The spot labeled "Discharge date" has been left blank. I think of another sign, the one that says "MOON ROCK" at the Smithsonian. Despite this sign, people frequently walk right by the moon rock, without noticing. I wonder what I am not noticing.

....

Because of my job as a science reporter, I've seen a lot of moon rocks, and meteorites, too. I've been inside the lab at Johnson Space Center that contains Apollo lunar samples. I've visited the back rooms at the Smithsonian's National Museum of Natural History, which houses the National Meteorite Collection.

It all mostly looks like a bunch of gray rocks.

Basalt, for example, is a dark gray, volcanic rock. Both the Moon and the Earth have tons of it. "The key chemical difference is that most terrestrial basalts contain some water, either dissolved in the glassy part or at very low concentrations in the minerals. Lunar samples are pretty much bone dry," geologist Darby Dyar told me when I visited her lab at Mount Holyoke College.

Together we looked at samples of lunar rock, brought home by astronauts, that she'd been analyzing. I held one lunar pebble in the palm of my hand. It was smaller than a dime, almost black, and slightly glittery.

"Here, this will be fun, let's get Mars," said Dyar. She pulled a fragment of a Martian meteorite out of a drawer and put that on my palm as well, right next to the bit of Moon. The Mars rock was an ever-so-slightly greenish gray.

Holding these scraps of the Moon and Mars in my hand,

I didn't feel the wonder that I'd expected and that I assume Dyar expected of me. These pebbles looked like rocks I could find just by walking outside the lab and searching under a bush. The strongest emotion I felt was a burst of love and affection for the ordinary gravel outside. If you could transport one of those humdrum stones to a museum on another planet and put up a sign that said "EARTH ROCK," some glorpy extraterrestrial would reach out with one of its many appendages to caress it tenderly, feeling awe at the universe.

I asked Dyar if the rocks recently collected by the Perseverance rover on Mars, the pristine samples sealed in metal tubes that were waiting to be brought home, would be just as boring-looking as the ones in my palm. Sure, she cheerfully agreed. Even though scientists would fight over each and every Martian mineral grain, she said, "they're going to look just like the stuff in my driveway. The only thing that makes them special is that they come from someplace else."

Then she added, "The cool part is that they're *not* special. There's uniformitarianism of geologic processes, which, to me, is also beautiful."

Uniformitarianism is the notion that the natural laws and processes at work today are at work everywhere in the universe and that they have operated in the past in the same way that they do today and will in the future. That was a radical idea to scientists who used to believe in the biblical story of creation, and it's a relatively recent one that traces back to British geologist James Hutton in the late eighteenth century, who wrote evocatively that in geology, "we find no vestige of a beginning, no prospect of an end."

I asked Dyar why so many rocks from the Earth and the

Moon and everywhere else are so resolutely gray—what element gives them this color? "Iron, mostly," she said.

That made me think of a remark made by another geologist, Oliver Tschauner, who studies minerals that lie deep within the Earth. He once told me that the core of our planet is mostly metallic iron and the inner core is solid, making it arguably the biggest gray rock around. "A piece of meteorite iron gives a good approximation for how it would look," he said, "if we could cool it to ambient conditions." But that's impossible. The giant lump of iron miles beneath our feet is actually unimaginably hot—over 8,000 degrees Fahrenheit. "At its place in the center of the Earth," Tschauner told me, "it would emit incandescent light like the sun."

Sometimes I think about the unseeable, sunlike center of the world as I reach up and touch the dark hunk of meteoric iron hanging from around my neck.

I call my father on his birthday. He is at home. To my great relief, he has somehow made it out of the hospital after spending more than two weeks there. He spent another couple of weeks in a rehab facility, got well enough to ditch the oxygen, and did enough physical therapy to start walking fairly steadily again.

"Happy birthday!" I say, asking if he and my mother are doing anything to mark the occasion. With enthusiasm, he says they are gathering friends and neighbors for a celebratory dinner at Alfie's and that I am invited. I can't make it that night, but I promise to come visit with the kids in a few weeks.

"I didn't get you a birthday present," I say, although I almost never do. "Do you remember that I got you a meteorite for Christmas?"

"Oh, yes," he says, politely. I don't know if he remembers.

I tell him that I'm going to try to collect some meteorite dust on my roof, and if I find any, I'll give it to him.

"That sounds like fun," my Dad says, again polite but with none of the gusto he had when talking about going to Alfie's. He hands the phone to my mother.

I didn't expect him to get excited about meteorite dust. Most people don't care about dust, even though "all are from the dust, and to dust all return," as Ecclesiastes says. But to me, meteor dust seemed like the only way I'd ever personally get to experience the discovery of something otherworldly.

That night was the peak of the Lyrids, one of the great annual meteor showers that occur when our planet's trusty orbit takes it through clouds of debris left by asteroids or comets that follow their own well-worn paths. Records of people watching this particular meteor shower date back to 687 BC. Light pollution would prevent me from seeing the "shooting stars" created when tiny rocks hit the atmosphere at over 100,000 miles per hour and flare up from the friction. But any debris that survived would fall as dust, and this kind of cosmic dust is by far the most plentiful source of extraterrestrial material here on our planet. Scientists estimate that some 5,200 tons of outer space dust reaches the surface each year; that's 14 tons per day, about the weight of three ambulances, drifting invisibly down.

After I hang up the phone, I clean out a plastic tub I've been storing fossils in. (They were ancient bivalves entombed in gray slate that I'd collected from the side of the road by smashing rocks with a hammer in the rain as my bored children waited in the car, occasionally cracking the window to yell, "How long until we go to McDonald's?") I fill the tub with about 5 inches of water. Then I put it outside my office window on the back roof, beside the solar panels.

As I type away at my computer over the next few days, I sometimes look out and see the plastic bin, slightly glowing in the sunlight. It makes me feel an emotional kinship with Harvey Nininger, who is sometimes called the father of modern meteorite studies. He was born in 1887 and raised in a farming home that had two books, the Bible and the Montgomery Ward catalog. When he was a kid, "meteors were regarded in about the same light as ghosts and dragons: mentioned rarely and never discussed seriously," he explained in his autobiography, *Find a Falling Star*. That's because even after scientists finally understood that the fireballs seen in the sky meant that certain rocks truly came from outer space, these events were seen as rare. The lumps of alien stuff were just curiosities. Oh, the biggest ones, impressive masses of iron that weighed dozens of tons, did get attention from collectors and museums; the American Museum of Natural History in New York, for example, features giant meteorites from Greenland that were hauled off by explorer Robert Peary, who was apparently indifferent to the meteorites' value to the local Indigenous people. But no scientists were making a career of going out and systematically *looking* for meteorites.

That's what Nininger wanted to do. In 1923, he was teaching biology at a small college town in Kansas when he read an article about meteorites. Only a few months before, he'd visited Chicago's Field Museum and seen meteorites on display. And as he described it, one evening after reading that article, just as he left the college chapel, "a blazing stream of fire pierced the sky, lighting the landscape as though Nature had pressed a giant electric switch. The blade of light vanished with equal suddenness, leaving a darkness seeming thicker than before." Immediately, Nininger bent over and made a mark on the sidewalk to plot the course of what he knew must have been a meteor. He also rushed to contact the state's main newspapers, urging anyone who saw the fireball to send him some key information. A confusing assortment of reports poured in. Nininger gamely sorted through them, looking for useful tidbits. He figured the space rock likely came down near the towns of Coldwater and Greenburg.

Over the next couple of years, he visited these towns frequently, speaking to schools and reporters and anyone who would listen. On one occasion, a church let him talk to the congregation for about three minutes. "I then joined in the order of worship," he wrote. After the service, a deacon told Nininger that he had a strange rock in his backyard, and it turned out to be "a forty-one-pound meteorite, an oxidized nickel-iron, that he had plowed up four years earlier. He said he had always wondered what such a heavy rock was doing in a field all by itself." The thrill of finding that meteorite, wrote Nininger, "can never be described." And a little later on, another farmer plowed up an odd rock—which turned out to be an 11-pound stony meteorite—and took it to the local newspaper editor, who called Nininger.

If meteorites were supposedly scarce, thought Nininger, how had he managed to find two good specimens so quickly?

Beginner's luck? Or were meteorites buried beneath the dirt all over the place, and folks didn't realize their significance? He started traveling rural roads, giving talks and passing meteorites around so that people could see and touch them. "Sometimes I would speak to groups of laborers coming off their shifts at mines or on road construction jobs. . . . More than once I walked out into the middle of a field where workers were hoeing cotton to show them a meteorite and urge them to keep eyes open for something like it," he recalled.

He envisioned a real "science of meteorites," one based on an abundance of samples, found with the help of curious everyday people—people who wanted to make a buck by selling their finds, sure, but who also must have appreciated someone with faith in their inherent capacity to appreciate and recognize artifacts from the world beyond. At that time, the nation had only a couple of geologists who knew much of anything about meteorites. One of them, the head of geology at what is now the Smithsonian Institution, wasn't encouraging. He told Nininger, "Young man, if we gave you all the money your program required and you spent the rest of your life doing what you propose, you might find *one* meteorite." Nininger ignored him, quit teaching, and devoted the rest of his life to finding and studying meteorites, as well as selling them to support his efforts. When he died at the age of ninety-nine, one obituary noted that he had turned up specimens from 226 meteorite falls that were previously unknown to science. "What else is there that can give one such a sense of wonder," Nininger asked, "as holding in one's hand an object that has come from space?"

I think of Nininger often during my own micrometeorite

search, filled with admiration—and jealousy—for his determination. Because I soon learn that finding even just a speck of meteoric dust isn't going to be as simple as putting out a plastic bin during a meteor shower, despite the oh-so-easy instructions on the Internet. One estimate I see suggests that every square meter of ground gets hit by one, or maybe two, micrometeorites each year. Even if I'm lucky enough to have my plastic bin positioned out there to catch one at the right time, these spheres of molten-and-then-solidified space rock typically are only a few hundredths of an inch across. A micrometeorite could fit in the valley between two fingerprint ridges.

Still, I go through the water in my bin with a strong magnet, to fish out anything that might contain iron. Then I use my son's microscope to examine the crud. I see jagged bits of mineral crystals that are clear and green and amber. What I am looking for is a minuscule, spherical version of the dark gray meteorite hanging from my neck.

I find nothing like that in the bin. Undaunted, I creep out onto my roof, lie down on my belly, and collect fine dust that has accumulated in the rain gutter. I run it through a sieve, then use the magnet again to select a smaller portion, enough to spread over a white dinner plate. My children watch me as I sit at the kitchen table, staring at the computer screen that displays the dark speckles under the microscope. My twelve-year-old son declares that my quest is "stupid."

"Come on," says my husband, "if your mother does find a micrometeorite, you know you'll love it. You'll want to name it! We'll call it 'Meaty.' "

"What's stupid about this?" I ask.

"You're literally microscopically inspecting gutter crud," says my son, "looking for an outer space rock."

"Well, yes," I say. "What's your point?"

"What's yours?" he retorts.

I don't have an answer. Except that the universe and everything in it that is *not me* always seemed so separate and unreachable, so much *out there*. This micrometeorite search feels like an ineffable chance to grab hold of it, or at least a small piece of it. I think of a short exchange I had with my daughter when she was about three years old, when I mentioned offhandedly that she was part of the universe. She frowned, saying, "No, I'm not, I am *in* the universe."

"You can be both in it and part of it," I told her, "just like your heart is inside of you, but also part of you."

"No," she said flatly. "You're wrong."

Who was I to argue? At my core, I felt the same.

And so I spend hours dragging the microscope over irregular shapes and pale colors that all jumble together, in between tending to mundane tasks like doing the laundry and feeding our tortoise and opening the mail. (One envelope from my parents contains a copy of my father's last will and testament, sent without any comment or note. A few days before it arrived, my brother told me that our father had had another fall, this time in my parents' own driveway, requiring a trip to the emergency room and stitches for a gash on his face.) Sometimes, in the dust, I see a dot that looks like a gray sphere and I get excited. Then, under magnification, it transforms into just another amorphous mass of mineral debris. I begin to despair. I can't believe that professional science educators act as though seeking micrometeorites is a fun science fair project for children. Here I am, a middle-aged science reporter with a quarter century's experience in conversing with scientists about their research, plus a *masters degree*, and I feel like this hunt might emotionally wreck me. Nininger would not be impressed. My daughter comes over to check on me one night as I sit at the kitchen table with the microscope. "I don't know, honey," I tell her. "It's not looking good."

"It might take a long, long time," she says, sympathetically. "Like, a year."

Occasionally, looking through the dust, I think of Walt Whitman's *Leaves of Grass*, and in particular the poem "Year of Meteors (1859–60)." It refers to "a strange huge meteor procession, dazzling and clear, shooting over our heads":

> *Year of comets and meteors transient and strange!—lo!*
> *even here, one equally transient and strange!*
> *As I flit through you hastily, soon to fall and be gone,*
> *what is this book,*
> *What am I myself but one of your meteors?*

For decades, no one understood what "meteors" Whitman had been talking about. Then astronomers saw a painting by Frederick Church called *The Meteor of 1860* and got inspired to hunt through old newspapers. They found reports that on the night of July 20, 1860, a meteor grazed the atmosphere and broke up, creating a trail of fireballs that traveled together across the sky, looking like balls of light shot from a Roman candle. This, at the time, was a major news event, but it soon was utterly forgotten, subsumed into the chaos of the Civil War. If I had been alive and working as a reporter in 1860, I have no doubt I would have been assigned to cover this meteor, just like I came to work on the morning of February 15, 2013, and had to crank out a quick report on the 60-foot Chelyabinsk meteor that had hit the atmosphere over Russia, exploding with a sudden blast and glowing like a second sun in the sky. The shock wave set off car alarms, shattered windows, and damaged thousands of buildings, injuring lots of people who had been going about their daily routines without any warning that they were about to experience the biggest meteorite impact in more than a hundred years.

One day, in my close examination of dust from my gutter, I see something that appears to be a metal sphere, and I grab a toothpick to try to tease it away from the rest of the mess. I'm startled when the toothpick comes shaking into the microscope's view like a battering ram, looking like a tree trunk stripped of its branches. I realize that my intense focus on micrometeorites has thrown my sense of scale out of whack. The dust looms large on my computer screen and in my mind, but each grain is so barely there that to the naked eye a pile of it looks like powder. Clumsily poking around with the seemingly enormous toothpick, I lose the spherical shape I'd been trying to take hold of and never manage to find it again. It most likely wasn't a micrometeorite anyway. My husband gives me a pep talk: "Who knows? The next piece of crud could be the one!"

But I start to think about the fact that even if I spot another sphere and manage to preserve it, I have no way of knowing if it is extraterrestrial or just an "industrial spherule" or "rounded mineral grain," the kind of thing described by Jon

Larsen, a Norwegian jazz musician and self-taught micrometeorite expert who pioneered the hunt for them in cities. His book, *In Search of Stardust*, features about a dozen pages of close-up photos of micrometeorites . . . followed by many more pages of nearly identical-looking objects found on roofs that were *not* micrometeorites. In one edition of this book, he says that it "is easy to check" if a sphere is extraterrestrial by looking at its elemental makeup with energy dispersive X-ray analysis. In a perhaps more optimistic version of this chapter in a subsequent edition, he writes that with enough experience, it is possible to tell the difference just by eye. *Maybe it's possible for you*, I think, feeling bleak.

One night at dinner, I ask my family if they think I should keep searching.

"Do you *want* a micrometeorite?" my daughter asks.

"I *thought* I did," I say. "Now I'm not sure."

"What would you do with it?" asks my son. I tell him I'd put it on sticky tape so as not to lose it, and then keep it in some kind of plastic specimen box.

"Imagine if other dust got in there," says my daughter, wryly.

"You mean, other dust would get in, and I wouldn't be able to tell the difference?" I ask, and she nods, grinning.

My husband says, "Imagine if you finally find one, and you're like, 'Hey, kids! Come look!' and just then *Boom!* A huge meteorite comes crashing through the ceiling." In his kindly vision, this fantasy meteorite does not strike me dead, but that's immediately where my mind goes.

"Or imagine you're like, 'Hey, kids!' And then a breeze comes blowing through," says my daughter, gesturing toward the screened door behind me and raising her eyebrows dramatically.

Sitting there at the table, being affectionately mocked with increasingly ironic meteorite discovery scenarios while eating the waffles my husband has made, I feel my shoulders relax. I know that I can try other ways of searching, like putting a magnet in our gutter to attract stray space dust, or asking a neighbor to give me all the crap that's collected in the bottom of his rain barrel. But I no longer feel any pressure. Maybe I'll find a meteorite, and maybe I won't. Maybe all I'll ever do is quietly sift through a bunch of ordinary, sometimes beautiful stuff, searching for something ethereal that I'm not equipped to recognize and probably won't ever truly understand.

....

The ocean sighs into the darkness as I watch the moonglade shimmer. Bright circles of light run across the sand as my children, flashlights in hand, search for skittering ghost crabs. We come to this beach every year for our summer vacation, and often we see shooting stars. This time, though, we're too early for the Perseid meteor shower, and the nearly full moon is so bright that I don't think we'll see any stray meteorites. I scan the heavens anyway, listening to the waves. I'm thinking about a phone call with my parents earlier that day; my father told me that my mother wasn't doing well, and when he handed the phone to her I learned that she'd had a fall in the kitchen. They'd spent hours in the hospital getting her checked out. Even though nothing was broken, the pain made it hard for her to walk, and she didn't think they'd make it to the memorial service being held for my aunt the next week. Gazing up at all the points of light in the black sky, I look for one with the steady movement of a satellite. This search reminds me of how it felt to run my eyes over microscopic speckles from my rooftop, hoping to glimpse another kind of gray sphere. A few times I think I see the thin, quick trail of a meteorite; each time, I know it is more likely just a glint of light reflected off my glasses, but there's no way to tell. My daughter begins sobbing, distraught over a crab she caught, "Crabby," who'd escaped and vanished into the dark before she got a chance to say a proper goodbye. She is nine years old, the same age my father was when the famous Sikhote-Alin meteorite slammed into the sky above Russia, blasting into fractured bits like the one I wear close to my heart.

Tonight, though, I'm wearing a different necklace, one that my daughter gave me. Like my father, I'm hard to shop for, but I love this necklace because she made it out of a stone she found here at this New Jersey beach. Cape May Point is famous for its "Cape May Diamonds," smooth round pebbles of almost pure quartz. I remember my daughter asking me, "Are these *real* diamonds?" and I had to say no, that quartz was the second most common mineral in the Earth's crust. I pointed out that these stones were still pretty, like beach glass. When wet, they seem to glow. I liked them even better than diamonds, I told her, because we could find them ourselves. She collected a Ziplock bag of beach pebbles, a mix of translucent, white, and pink, and I bought her some pendant chains and wire so that she could make jewelry. The necklace I'm wearing doesn't feature one of the completely clear "diamonds." Instead, this stone is a small, flat oval that's a pale silver-white, with almost imperceptible veins of gray that I assume are iron. It reminds me of the Moon. My daughter has wrapped the pebble, too loosely, in clumsy loops of thin silver wire. Part of me is scared to wear her gift because I know this rock could easily slip out. In fact, it's not just a possibility, but a certainty. Inevitable. And I know that when that happens, the stone will be lost forever. It will be gone so quickly, with no chance that I'll hear the too-faint whoosh of its fall—or the silent thunk as it hits the ground.

A very charming young black hole

I had my first kiss—or something I called that at the time—when I was twelve years old. It happened in the crazed Lincoln Log cathedral that is the Old Faithful Inn. The inn bills itself as the largest log structure in the world, with wood beams and twisted branches that create stairs and balconies that rise nearly 100 feet, all the way up to a platform called the "Crow's Nest," which supposedly fulfilled a tree house fantasy from the architect's childhood. I couldn't go up there. The final set of stairs that led to it was roped off, and a sign said this was because an earthquake had damaged its structural integrity.

Which was strange, because danger didn't seem to put anywhere else off-limits in Yellowstone National Park. As my parents took me around on family vacation excursions, I studied the ubiquitous signs: "BEAR ATTACK! Are you prepared to avoid one?" and "UNSTABLE GROUND—BOILING WATER." Any time I left the inn and walked out over the moonscape of the geyser basin, I was sure to encounter bison. The alarmingly large, shaggy beasts roamed parking lots and pathways with impunity, impassively chewing grass right next to signs that said "WARNING—MANY VISITORS HAVE BEEN GORED BY BUFFALO. These animals may appear tame but are wild, unpredictable, and dangerous." Then I read that the entire Yellowstone region was on top of an ancient supervolcano just waiting to go boom and spew lava 40 miles in every direction. Such an eruption hadn't occurred for hundreds of thousands of years, but still. Unknown violence seethed beneath the surface. You could smell it burbling up in the mud at Sulphur Cauldron and see it in the sapphire blue of the Grand Prismatic Spring, which simmered relentlessly, shrouded in billowing steam, ready to boil you alive. I would wait for hours beside Grand Geyser, the world's tallest predictable geyser, ostensibly reading a book while distracted by fantasies of what might be going on inside the Earth below, the mysterious buildup of pressure and heat. Then I'd hear a sudden wet gurgle and a ranger shouting something like, "There she blows!"

After a day of pondering these disturbing natural wonders, we'd have dinner at the inn and my parents would go do—what? I'm not sure. Most likely have a drink while sitting out on one of the wooden balconies, listen to the nightly pianist play old show tunes like "Try to Remember," then go up to our hotel room to sleep. Meanwhile, I wandered around, unsupervised, until all hours of the night. I had no curfew, despite not even being a teenager.

One evening I was sitting alone in one of the leather chairs in front of the inn's central fireplace, an imposing 85-foot tower of stone. The metal screen in front of the blazing fire was decorated with an erupting geyser; hanging above the fire was a giant clock. It was late enough that the bused-in tourists and dinner guests had all cleared out. I had a writing tablet with a picture of a windmill on the cover. On one of its pages, I was scribbling out some words I had written, circling over them with my ballpoint pen again and again until the whiteness of the page had nothing left but several big black circles. I saw a group of college-aged guys pass the reception desk and head toward the Bear Pit Lounge. One of them had a can of beer, and I heard him tell his buddies that he'd finish his drink outside rather than take it into the bar. "It's more than half a can, so it might be a while," he told them. He sat down next to me. He was just a normal guy: short blonde-brown hair, brown eyes, T-shirt, jeans. He sat drinking his beer and looking at the fire for a moment, then leaned over to ask what I was drawing. We gazed together at the black circles.

"I don't know," I said.

He asked if he could add to them, and took my pen from my hand. He drew a geological diagram, showing the last volcanic super-explosion. He labeled rock layers: igneous, limestone, sandstone. As he sketched, he said it was a mystery as to why volcanic ashes created certain rock features on one side of Yellowstone and not the other. I pointed to one of my dark circles of ink and noted that in that other direction, there appeared to be a large black hole.

The words "black hole" came to my lips even though I didn't really know what they meant. A black hole was something terrible that pulled you in, something so powerful that not even light could escape. All that I had gleaned from the 1979 sci-fi thriller *The Black Hole*, which I saw in a movie theater with my family. It seems like an inappropriate film for a kindergartner, but perhaps my parents thought that anything made by Disney would be harmless. *The Black Hole* has been called the most scientifically inaccurate movie of all time, but it nonetheless made quite an impression on my young mind.

My childhood happened just as the science of black holes began to grow and mature; my coming of age was entangled with its coming of age. Oh, the idea of a "dark star" had been around since 1783, when a small-town English rector named John Mitchell imagined a star so massive that it pulled in its own light. He suggested that the universe might be hiding many dark stars, whose very nature rendered them invisible. More modern scientists, including Robert Oppenheimer, applied Albert Einstein's theory of general relativity to the notion and came to similar conclusions. But the whole thing sounded so nuts that for a long time most physicists didn't take it seriously. Physicist Robert Wald has said that this subject was either ignored or deliberately avoided, and if someone tried to talk about black holes in a professional setting, "you'd get various combinations of wry smiles or sneers, depending on the personality of the people in the audience. And when you were done, it was clear the reaction was: 'Ok, that was fun, now we've got to go back and do some actual physics.' " Other than general relativists like himself, Wald recalled in a lecture, "black holes were not generally taken seriously by 'hard core' astrophysicists and physicists until about the mid-1970s."

I was born in 1974. And when I sat next to this strange man in the lobby of Old Faithful Inn, blurting out the words "black hole," it was 1986. I wasn't a child, but I wasn't an adult, either. The same could be said about the science of black holes. Astronomers had started to make weird observations that could be explained by hungry, unseen companions lurking near certain stars. "Until about the mid-1980s, one never spoke about the presence of black holes in astrophysical systems, just 'black hole candidates,'" noted Wald, but as the evidence grew, scientists began to feel that it was safe to drop the word "candidate." Black holes existed. They were real. Still, even the best scientists in the world could say very little about them for sure. Two months after my preteen Yellowstone trip, the prominent journal *Science* featured an article entitled "The Galactic Center: Is it a Massive Black Hole?" The answer, we now know, is yes. The heart of our Milky Way is a black hole that's about four million times more massive than our sun, and it seems to be surrounded by thousands of smaller black holes. Researchers think a black hole lies at the center of most, if not all, galaxies. But back then, with no conclusive observations to offer as proof, physicist Fred K. Y. Lo could only venture that "the extraordinary phenomena" within the central light year of the galaxy "definitely call for something unusual."

I didn't read that article in *Science*. I was twelve; I probably didn't even know the journal existed. I knew almost nothing about black holes or, for that matter, the rest of the universe, including myself. What I did know is how to flirt with a much older man. Just after I teasingly pointed to the "black hole" that was altering the deposit of supervolcanic ash in his geology diagram, one of his friends came out of the bar to check on him. My new companion shared our theory: a black hole at Yellowstone influenced its mineral deposits. His friend laughed and said, "Excellent! Let's write a paper." I told him I'd better get proper credit. As more of his friends came out from the bar, the geology-drawing guy asked my name—for our joint scientific publication. I told him to guess and, when he said "Sacagawea," I feigned astonishment and pretended that he was the only person who had ever guessed my name correctly. Then he grabbed my notebook and wrote, so his friends wouldn't overhear:

I WANT TO KNOW YOUR NAME PLEASE. HOW OLD ARE YOU?

I wrote back: *How old do I look?*

SEVENTEEN TO NINETEEN

HA! Yep 17. I look younger tho.

As a middle-aged woman, I wonder whether any sane adult man could plausibly mistake a twelve-year-old girl for a seventeen-year-old. I don't believe that he believed me. It seems impossible, although, in fairness, most twelve-year-old girls are not hanging around hotel lobbies by themselves late at night, even though this was the 1980s, when a whole generation of kids was largely left to their own devices. His friends stood around for a bit, while we ignored them, until one of them finally said, "Well, the Chief seems to have things under control here," and they walked off. Meanwhile, the two of us were writing back and forth to each other, fingers touching as we passed the pen.

DO YOU HAVE A BOYFRIEND?

Sometimes

WRONG TELL ME THE TRUTH YOU COMPULSIVE LIAR

I told the truth. Sometimes I do.

YOU ARE TOO PRETTY TO SOMETIMES HAVE A BOYFRIEND. HAVE A ♡ AND TELL ME THE TRUTH

What the hell are you talking about? I had one, we broke up. Funny old thing, life.

WHY DID YOU BREAK UP?

Because he cheated on me.

OH WHAT A DICK WHY WOOD HE WANT TO DO THAT TO YOU.

Because I'm not a slut like her.

This whole exchange makes me cringe. An adolescent calling life a "funny old thing" seems so transparently pathetic; and yet, even as I wince, I realize that this is exactly the same self-criticism that made my child-self adopt that pose in the first place. *How I was wasn't good enough,* I thought back then and, apparently, now, because I feel none of the tenderness that seizes me today whenever I pass by random kids on the sidewalk who are saying "fuck" or smoking cigarettes in a desperate attempt to seem jaded. I note with a little sadness that the immediate fictitious narrative created by my younger self revolved around me being betrayed. And it comes as a shock to see myself unironically calling someone—even an imaginary someone—a "slut," a word whose meaning I seemingly accepted and used without irony. I console myself with the thought that perhaps the "slut" comment was meant as a self-defensive move, as if to say to this unfamiliar man: Don't try anything with me, because I am not that kind of girl. His response was immediate:

THANK GOD! YOU ARE A VERY CHARMING YOUNG ~~BLACK HOLE~~ GIRL.

He wrote "black hole" teasingly, then crossed it out. No one had ever called me a black hole before. I had no idea, back then, where the term even came from. Most people don't know. The astronomer John Wheeler generally gets credit for "coining" this formulation, but that doesn't seem to be true.

In his autobiography, Wheeler claimed that in 1967, he gave a talk in New York. "I remarked that one couldn't keep saying 'gravitationally completely collapsed object' over and over," he wrote. "One needed a shorter descriptive phrase. 'How about black hole?' asked someone in the audience. I had been searching for just the right term for months, mulling it over in bed, in the bathtub, in my car, wherever I had quiet moments. Suddenly, this name seemed exactly right."

He wrote this even though Wheeler's colleague at Princeton University, Robert Dicke, had been regularly using this term for years. Dicke had been making a metaphorical link between this celestial object and the "black hole of Calcutta," a prison that became notorious during Britain's rule over India. It was a small prison, but soldiers reportedly once shoved so many prisoners into the cramped space that over a hundred men were either suffocated or crushed to death. A physicist named Carlos Herdeiro, along with colleague José Lemos, has pointed out that in 1996, Wheeler told another scientist that he and Dicke had privately used this term together. "It is intriguing that John Wheeler suppressed, in his description of the origin of the term black hole, the former practices of this terminology," noted Herdeiro and Lemos in an article they wrote on the genesis of the name. "It is impossible that Wheeler and Dicke, as colleagues at Princeton, never touched upon the subject or that Wheeler had never heard such an idiosyncratic expression as 'black hole of Calcutta' in the Princeton corridors." One of their fellow scientists, who organized the event where Wheeler supposedly seized upon the term, looked visibly uncomfortable when asked about the account in Wheeler's autobiography, and said, "Wheeler could have told the story he wanted."

What is undeniable is that Wheeler dropped the term into a 1968 publication, giving it gravitas and resulting in its widespread adoption. In an interview, Wheeler once said that he latched onto this name as "an act of desperation" because he wanted to force people to believe that these bizarre objects existed. He decided to start casually using "black hole" in his talks and writings, as if "it were an old familiar friend," he wrote. He wondered if it would take off, and it did: "By now every schoolchild has heard the term."

Not everyone reacted to it in the same way. Wheeler has written that physicist Richard Feynman thought the term was "suggestive" and "he accused me of being naughty." At one point, Feynman even said that Wheeler was being obscene. That was when Wheeler made an analogy about the way that black holes could be described by just three basic properties—their mass, charge, and angular momentum—and remarked that "a black hole has no hair." What Wheeler meant was that black holes hide their secrets so well that they have no external features that could be used to tell one from another, like people with heads so bald that no hairstylist could ever give them a distinctive look. Scientists call this the "no-hair conjecture." Feynman didn't think the idea of a hairless black hole was fit for polite company, recalled Wheeler, who remarked, "I guess Dick Feynman and I had different images in mind."

Feynman wasn't the only one who felt this way, however. In 1969, the editor of the journal *Physical Review* refused to publish a technical paper that used the "no-hair" idea, saying he would not print obscenities, according to Wheeler's student Kip Thorne, who doubted his mentor's protestations of innocence. "It is hard for most of Wheeler's colleagues to believe that this conservative, highly proper man was aware of his phrase's prurient interpretation," Thorne wrote in his autobiography. "I suspect otherwise; I have seen his impish streak, in private, on rare occasion."

Wheeler has said that to him, "black hole" just felt right because the words echoed another well-known scientific term, a "black body." In physics, this is an idealized body that absorbs all the radiation that falls upon it and emits radiation at the maximum rate possible. "The black body is a perfect absorber, and as perfect an emitter, as it is possible to be," wrote Wheeler. "A black hole has one of these characteristics, but not the other. It absorbs everything that falls upon it. It emits nothing."

My own reaction to the words "black hole," written on my notebook by a man's hand, was to smile. It sounded adult and mysterious, strong and alluring. I felt less enthused about the "young girl" part of his compliment, so I wrote back:

You sound like my mother. How old R U?

YOU SOUND LIKE MY UNCLE ON MY THIRD COUSINS SIDE. 20.

He then proceeded to write down his full name—let's call him John—the name of the college he went to, and the name of the place where he grew up. "If you lived there you would be the most gorgeous girl in town," he wrote. He begged me to tell him my name, and I lied, saying my name was Beth. I don't think he lied, because now, more than three decades later, when I take the personal information he gave me and type it into an Internet search engine, I find a man with that name, from that town, who would have been twenty-two when we met. That, added to some other matching details he told me, makes me think that this guy could be the black hole romancer. If so, then the only lie he told was to make himself two years younger.

It would be so easy to contact this stranger who briefly passed through my sphere of influence. I could call him up and ask what, exactly, he was thinking. The trouble is, I suspect that our game of lies would pick up again as if our time apart had never happened. The only difference is that this time the two of us would switch sides, and we'd circle around and around the truth without getting any closer. I perform an experiment; I close my eyes and focus my entire being on this man. I wait for an emotion. Instead, I feel a void: complete indifference.

The back-and-forth scribbling we were doing on my notebook, a flirtation through text, may seem unlikely. To my twelve-year-old self, however, creating an alternative persona while writing to a man felt appallingly familiar. I did it all the time—through the computer. Home computers were new in the mid-1980s, and my older brother introduced me to the world of bulletin board systems (BBSs). These were primitive electronic forums that people hosted on their personal computers, which they hooked up to their house phone line so that other computers could dial in. The popular ones often had a busy signal when I'd try; I'd have to redial again and again until finally the squawk of the modem announced the connection. A welcome message would scroll up on my screen, often a campy illustration made out of ASCII text. I'd read the postings in the public forums, check my personal messages, then chat with the system operator—some guy sitting in front of the computer that my computer was connected to through the phone.

My parents had given me a dedicated phone line and encouraged my interest in technology. What they didn't realize was that they'd effectively installed a direct route for strangers to get into my bedroom. It didn't occur to anyone to worry about child predators online; instead, back then, it was the kids that were the threat, with law enforcement focused on reckless teenage hackers. And, indeed—*although I am not saying I did this*—it was easy to get illegally obtained information, such as calling-card numbers so you could call long-distance BBSs without paying the phone company (or without your parents knowing). It was also possible to ask your anonymous BBS pals for their phone numbers, which I did, so I could talk with them and hear their voices and let them hear mine.

And I lied. I lied constantly. My modus operandi was to find out what the guy wanted and then say I was that. A friend of mine recalls that we first spoke together when he was fifteen, and he was thrilled to actually be talking to a real, live girl his own age. I'd told him I was fourteen. After he became friends with my brother, he learned that I had been eleven. Another guy used to call me from what he said was his job at a morgue. I have no idea if he actually worked at a morgue. He was seriously into the singer Sheena Easton, so I said I was too and had to fake my way through conversations about her music. I became an expert at making vague remarks, like an elderly person trying to hide the onset of dementia.

Despite all this experience, my lies in the Old Faithful Inn felt different, riskier. The man I was talking to was right there next to me, in his masculine body. Maybe it unnerved me to not be able to hang up the phone or log off the computer. Several times I told him I had to go soon—more lies. At one point, I told him I could only stay for another half hour.

O.K. DO YOU MIND IF I STAY?

No, but we look pretty god damn suspicious . . . a 20 yr old drunk and a 16 yr old basket-case.

I AM NOT A 20 YR OLD DRUNK. I AM A 20 YR OLD SLIGHT BUZZ. I KNOW PERFECTLY WELL WHAT I AM DOING. MY FRIENDS ARE PROBABLY DRUNK BUT I CUT OFF A FEW HRS AGO TO TALK TO YOU. IF I WOULDN'T HAVE STOPPED, I KNOW I WOOD BE DRUNK BY NOW. I HAVE A NIECE NAMED BETH. I HOPE SHE GROWS UP TO BE A MANKILLER LIKE U.

If he noticed that my supposed age had dropped by a year, he said nothing. He leaned over and whispered, "I forgot to write, you're not a basket-case, either." That's when I told him that my name wasn't Beth. "The truth, I trust you now," I wrote, telling him that my name was Nella, which was yet another lie. It was, however, what was on a name bracelet I was wearing that I'd bought in the inn's gift shop, as it was the closest one to my name that they had on the rack. I showed him the bracelet as proof of my newfound trust and honesty.

THAT IS A VERY PRETTY NAME AND NOTHING TO BE ASHAMED OF. WHO GAVE YOU THAT?

My ex boyfriend. A peace offering. I want to dump it in a geyser.

YOU ARE A GOOF. I WISH YOUR NAME WAS BETH BECAUSE I REALLY LIKE THAT NAME. NOT THAT I DON'T LIKE NELLA DON'T GET ME WRONG. YOU ARE A VERY SPECIAL PERSON AND I AM SURE THAT SOMEWHERE DOWN THE LINE A VERY HANDSOME AND CHARMING MAN WILL COME ALONG AND SWEEP YOU OFF YOUR FEET.

Maybe.

WHAT'S WITH THIS MAYBE STUFF? I KNOW IT FOR A FACT.

Tell me the evidence. We have to support this theory.

FIRST OF ALL, NELLA, YOU ARE A VERY BRIGHT AND ATTRACTIVE YOUNG LADY. I WISH I WERE FOUR YEARS YOUNGER. I DO NOT KNOW WHAT IS WRONG WITH YOUR EX-BOYFRIEND BECAUSE IF I HAD A CHANCE TO TAKE YOU OUT I WOULD GO IN A SECOND AND NOT LET GO. YOU ARE SUPER. BELIEVE ME I CAN TELL. TELL THAT GUY TO GET THE HELL OUT OF HERE.

"That guy" was a man who had come over and plopped down in a chair on the other side of me. He asked what I was writing. I told him that it was secret spy stuff, because me and John were Soviet spies. He laughed and asked if John was my brother. "No," I said, and turned away. I focused on writing a reply to John in the notebook. Then a whole group of inebriated merrymakers spilled into the lobby. The bar had kicked them out, and they tried to keep the party going by gathering around the chairs in front of the fireplace and its dying embers. When the guy next to me got up and wandered off, another man came and sat down in the chair. He began talking to me in a slurred way, leaning in close and breathing beer into my face. Then he started talking across me to John. On the notepad, I scribbled:

Do we no him?

NO! DO YOU?

No! We get rid of one and another comes.

MAYBE IT'S THAT GORGEOUS FACE OF YOURS

We got to get out of here. He scares me. Can we go to the 2nd floor?

I DON'T MIND. IF YOU WANT TO GO I AM GAME.

The reason I am worried is that the bar just closed and "my face" almost got me in trouble with some drunk fool who was old enough to be my dad. Anyway, I have to go soon.

R U TALKING ABOUT ME?

No. I am talking about DRUNKS.

This story about past trouble with a drunk was the only thing I told him that was true. My notebooks from that vacation mentioned some man in his thirties, with a beard and black hair, who came up to me and "made passes." I had compiled a list of all the guys I met that week and ranked them on a scale of one to ten. This person got the lowest ranking possible, so he must have creeped me out—which is saying something. Already, I had become accustomed to a constant swirl of male attention. I apparently exerted some force that drew them in—all I had to do was sit there. Even though I had assured John that I was not talking about him, my description of a much older drunk dude who was trouble clearly got his attention. He asked me, flat out:

WHAT DO YOU THINK OF ME? AM I A PIECE OF GARBAGE OR WHAT? EVEN THOUGH I AM A BUZZED UP CHARACTER.

You are not a piece of garbage. But you can't spell.

I AM JUST BEING GOOFY.

I will put YOU on the spot, what do you think of me? & I have to go soon.

YOU ARE A DEAF MUTE WHO HAS SOMETHING ABOUT BLACK HOLES.

He was right. I was deaf and mute, and I did have something about black holes. But I didn't know it then. I also didn't know that a few months after I was born, two physicists made a famous bet. Astronomers had observed a strange X-ray source in the constellation Cygnus and had pinpointed it to a blue star that, by itself, couldn't account for the X-ray emissions. One possible explanation was that the star had a hidden companion: a black hole surrounded by a violently swirling disk of gases, one in which matter was accelerating and heating up enough to spew out X-rays before reaching the point of no return and falling in. This was the first time anyone had spotted a potential real-life example of something that had been, up until then, a nutty theory spun by pure math. Physicists Stephen Hawking and Kip Thorne both thought that Cygnus X-1, as it was called, was truly a black hole. Just for fun, Hawking bet Thorne that it wasn't. They scrawled down the terms, and Thorne kept the paper in his office for years. Eventually, Hawking decided that enough evidence had accumulated to show that this celestial object was truly a black hole, and he took action. "Late one night in June 1990, while I was in Moscow working on research with Soviet colleagues, Stephen and an entourage of family, nurses, and friends broke into my office at Caltech, found the framed bet, and wrote a concessionary note on it with validation by Stephen's thumbprint," Thorne recalled, adding that his mother and wife were "mortified" by the bet's stakes.

Those stakes are sometimes genteelly referred to as "magazine subscriptions" by people who want to elide the fact that what Thorne had won was a bunch of pornography. "I have conceded the bet," Hawking wrote in his best-selling popular science book, *A Brief History of Time.* "I paid the specified penalty, which was a one-year subscription to *Penthouse*, to the outrage of Kip's liberated wife." But his wife told physicist Janna Levin that she hadn't actually been outraged: "My reaction was one of surprise, more than anything... because I thought that the women's movement was well underway in sensitizing folks about these things. Clearly I was mistaken." One journalist, writing for *Motherboard*, summed up the whole episode this way: "The first formal acknowledgement that black holes really exist in nature was sealed with a subscription to a softcore porn rag by two of the world's leading cosmologists."

When Hawking conceded this bet, I had just turned sixteen. I was finishing my junior year in high school, and that summer I started working as an intern in a university science lab. There I met a graduate student who was ten years older. We used to go together to get lunchtime takeout for our lab mates on his motorcycle. Motorcycle rides led to kissing. He told me that everyone at the lab would be furious if they found out, so we kept it a secret. He fondly called me "jailbait," but two months earlier I'd become legal under New Jersey law. I remember one night he had to tend to an experiment after hours, and I sat on his lap in a chair in front of a gas chromatograph, letting him unbutton my blouse to see the green-

and-black Victoria's Secret bra that I'd bought to wear for him. Back at his apartment, on the floor, our kissing progressed to sex and I went along with everything in silence even though I wasn't on the pill and he didn't wear a condom. I didn't tell him that I had been a virgin until later that night, when we were both in the shower and he asked me point-blank how many times I had done it. My answer seemed to astonish him. He claimed he never would have had sex with me if he had known. I don't know how he thought he could have known. He knew almost nothing about me. He should have at least understood what another twenty-something-year-old scientist had deduced almost immediately in the lobby of the Old Faithful Inn, the basic, most important thing that Wheeler had wanted to emphasize when he first seized on the words "black hole": It absorbs everything that falls upon it. It emits nothing.

Thanks a lot. I have to go home.

YOU ARE VERY PRETTY AND SOMEONE (I'M SURE VERY CUTE) WILL COME ALONG AND BE YOUR KNIGHT IN SHINING ARMOR. I DON'T THINK YOU HAVE ANYTHING TO WORRY ABOUT WHATEVER YOU DO DON'T GET IN A HURRY ABOUT ANYTHING CAUSE YOU ARE YOUNG. BUT ALWAYS REMEMBER THE GUY NAMED JOHN WHO STOPPED AND TALKED TO YOU IN THE LOBBY

P.S. CAN I WALK YOU HOME

There was no home. We were there at the inn. Perhaps he thought I was staying in one of the historic cottages outside or in the neighboring lodge. Maybe he assumed I had my own hotel room. It's possible he planned to accompany me as I walked through the darkness, to protect me from drunken louts—my knight in shining armor! Or maybe his plans were different.

But I was just going upstairs. I did not have far to go. Together we walked up a staircase made of halved logs and gnarled branches. The fake candles glowing on the wall were dim, and the thick timber banister felt cool and smooth as we climbed up and up. I could sense his body moving behind me. I was trembling. At the bottom of the next staircase we stopped. I turned around, standing on the step above his, bringing us face to face. There, in the almost-dark, something happened. I could not tell you what. All I know is that I turned and ran. I ran up the rest of the stairs to the next balcony and rushed across it, passing the rickety staircase that led to the dangerous Crow's Nest. If it hadn't been roped off, I might have gone up there, I was in such a panic. Instead, I got to the right hallway and reached the room I shared with my family. I quietly opened the unlocked door and crept inside. A loud snore came from the bed where my father and mother were sleeping.

Running away was the only thing I did that night that seems remotely adult. As the mother of two small children, I can state unequivocally that when a twelve-year-old is alone with a man in a dark corner of a hotel, way after midnight, and the man has had some beers and has spent hours writing intimate notes back and forth filled with words like "mankiller" and "gorgeous," the right response is to flee.

That was not what I thought then. I filled my notebook with regrets: "I acted so dumb when we kissed goodnite. He's 20 years old, he expects more than just a kiss. I wish I could see him again, I'd be a lot better. I LOVE HIM!!!" I sat in the lobby of the inn, hoping he'd come back, fantasizing about him picking me up to take me out on a fancy dinner date, with me wearing my mother's blue silk dress. And I wrote him letters, on the assumption that someday, somehow, I'd find his address. I wrote about how he'd made me laugh when I was lonely, and how he was perfect, and how I missed him: "I dream about what I should have done to make that night special for you. I know I am nothing to you, but maybe if we had more time together you could like me even if I'm not the most experienced, pretty, and exciting girl you ever met. I don't know why you talked to me, but thank you! You made my life a little brighter. . . . At the end of that night I acted like a stupid child and I'm sorry, but I've had some pretty bad experiences with older men. Please forgive me! I'll do anything to make it up to you. You are the best person I have met or will meet forever. I'm so sorry. Nella." I had learned an important lesson. The next time, I'd do better. The next time, and the next, and the next, I'd keep all my fear secret, along with hurt and sadness and shame and doubt and anything else that might stand in the way of my becoming whatever it was he wanted, which is what I assumed I must want to be.

The big mystery about black holes, these days, is whether what's locked inside is truly gone forever. It's called the black hole information paradox, and it comes from something that Hawking realized at about the same time he made that obscene bet. He mixed the rules of quantum physics with general relativity and determined that black holes must emit *some* particles. And in doing so, the black holes would gradually shrink away into nothingness, a process that's been compared to an aspirin dissolving in a glass of water. His report on this idea, published in *Nature* in 1974, the year I was born, has been called "the most beautiful paper in the history of physics." It raised the disturbing possibility that all the information trapped inside a black hole might disappear, an idea at odds with scientists' belief that information is fundamentally indestructible. Today's titans of theoretical physics argue about whether or not this is true and devise exotic theories to explain one fate or another. At times, before he died, Hawking himself struck an optimistic note. "Black holes ain't as black as they are painted. They are not the eternal prisons they were once thought," he once said in a speech. "Things can get out of a black hole, both to the outside, and possibly, to another universe. So, if you feel you are in a black hole, don't give up. There's a way out."

I wonder about that. In the notebook where this stranger and I had written messages together, after his final plea to always remember, I'd added a childish scribble: "First kiss! Forget him?! Never!!!" But I don't think we actually kissed. My lips have no felt memory of touching his. I create scenarios in my mind to see if any of them feel real. I kiss his cheek; I kiss the air beside his cheek; he asks for a kiss and I turn and run. Nothing seems right. The journalist in me—the compulsive truth-teller—wishes I could rely on the contemporaneous account. After all, the girl who was there said it was her first kiss. If anyone was in a position to know, it would be her. She even wrote it down for posterity, in black and white. The only problem is: that girl lied. She lied all the time, about everything, to everyone, not excluding herself. In the story she was writing about her life and what it meant, she crossed out her own words so many times that what she started with became completely obscured beneath a growing swirl of black. Sometimes I think it might be possible to stare at this darkness long enough to get through, to discern what was there at the start. Other times, I know that's just another lie.

Part II

Pum pum pum

My daughter considers her fist. She's been listening to me read a book called *The Heart,* which says, "Make a fist. This is about the size of your heart." This is a factoid I remember from my own childhood. My three-year-old holds her fist near her chest, trying to imagine the unseen heart within, as I read about how your heart beats nonstop for every second of your life. My daughter says, thoughtfully, "It stops when you go to sleep." Actually, her older brother informs her, it *never* stops. She says nothing but I sense she does not believe him.

This does not surprise me. Learning about the heart takes time, as this strange muscle squeezes and releases, squeezes and releases, over and over, thirty-five million times a year, more than two billion times in an average life span. But it's also true that we never really come to understand the heart. Even in midlife—perhaps especially now—pondering my own heart produces a queasy vertigo, like trying to wrap my mind around infinity. The other organs work in wet silence. When I think about, say, the spleen, I feel nothing. Yet for those who care to listen in, the heart makes its mechanical presence known as its valves open and close, open and close, pumping 6 quarts of blood through tubes that stretch thousands of miles, starting with your aorta, which is as thick as a garden hose.

Forget the thousands of years of symbolism, the Sacred Heart of Jesus, the weighing of the heart by Egyptian deities, and all the honky-tonk country songs about hearts that "broke." What matters is the fleshy truth of the heart, the thrum of something that's impossible to grasp.

At Thanksgiving, I reach inside the turkey carcass to retrieve the plastic bag, then root through the giblets until I find the bird's tough little heart. The dark red meat has gobs of fat and cutoff tubes sticking out, and I put it aside to show the children. My hope is that, bit by bit, I can help them get closer to understanding the heart, the real heart, not the abstract shape they cut from folded sheets of construction paper on Valentine's Day.

I hope this even while knowing my efforts are pointless. When I was a child in the 1970s, my parents took me and my brother to Philadelphia's Franklin Institute, where we clambered inside its famous model of a human heart, one big enough for a 220-foot-tall person. I vaguely recall being inside a red space, surrounded by whooshing. This enormous heart has been updated and modernized through various "surgeries" since its creation in 1953. When my own children crawl through this giant, fake heart—because I will take them on a pilgrimage—their cardiac journey will no doubt be more high-tech, though ultimately equally meaningless.

After all, how many hearts did I encounter when I was growing up? I cut open a worm, a frog, a fish, a fetal pig, and a cat. Once I took the cat corpse home on the school bus, to study its innards in the room above our garage. I remember its soggy fur, the smell of some formaldehyde-like chemical, and the unreal blue and red of its veins and arteries, which had been injected with latex dye. My own cat curled purring around my legs, then jumped on the table to sniff this obscure object, as my gloved hand touched another cat's inert heart. None of this made much of an impression. A motionless heart says nothing, while a heart that goes silent says everything.

For about three weeks, everyone lives without a heart. That's how long it takes for two tubes to form in a developing embryo and then merge into one tube that spontaneously starts beating, with blood coming in through the bottom and leaving through the top. The heart is the body's earliest functional organ. I did not know this the first time I was pregnant. At about six weeks in, I had some bleeding, and my doctor sent me off for a sonogram. I'd never had one before. A wand went inside me, and on the machine's monitor, the embryo looked like a kidney bean, with a slight fluttering in the middle. "There's the heart," the technician told me brightly, and then, to my surprise, she said, "Let's take a listen!" She flipped a few switches and the room filled with a rapid *pum pum pum*. "Sounds great!" she said, and printed out a small fuzzy photo for me to take home.

Unbeknownst to me at the time, what I'd heard wasn't a

"heartbeat" at all. When a doctor uses a stethoscope to listen to the heart of an adult or a child, the *pum pum pum* comes from the opening and closing of the heart's valves. "At six weeks of gestation, those valves don't exist. The flickering that we're seeing on the ultrasound that early in the development of the pregnancy is actually electrical activity, and the sound that you 'hear' is actually manufactured by the ultrasound machine," one obstetrician explained to a colleague of mine, who was reporting on efforts in Texas to ban abortion after a "fetal heartbeat" can be detected.

What I'd heard was an artificial creation; a heart like my own, with noisy chambers and valves, wouldn't develop for weeks. Nonetheless, the sound and associated flickering had a startling effect. Dazed, I staggered out of the exam room, clutching the photo. Up until that point, the pregnancy had seemed theoretical at best and then, after the bleeding episode, written off entirely. A few weeks later, my husband and I took a bus downtown to get a second sonogram. He was cheerful, full of excited anticipation to have the same experience that I'd told him about. "Look," I said, "You have to prepare yourself for the possibility that this is all going to go bad." He rolled his eyes. At the doctor's office, he stood at my side as I lay there with my shirt pulled up in the darkened room, with the ultrasound technician pushing the wand against my belly. This went on for a suspiciously long time. The technician finally said, "I'm sorry, but I'm not finding any heartbeat." I can't remember what, if anything, we said. There wasn't anything to say. I now know the name for this kind of pregnancy ending: a "silent miscarriage."

Perhaps it's not surprising that when I got pregnant again,

I became fixated on the sound of the heart. At each visit with my midwife, as my pregnancy progressed and my belly grew round, she would rub a bit of clear jelly on my skin and press down with her Doppler wand. A speaker crackled with static as she poked the wand this way and that, trying to catch the fetal heart's quick rhythm—one twice as fast as my own, like a rabbit's or a cat's. We never talked or joked during the search. I held my breath, waiting. Long before I knew whether my child was a boy or a girl, I knew the sound of its heart. And, in turn, this strange creature knew my own heartbeat, the constant low frequency rhythm of its dark, watery world.

Two days after my son was born, I even got to see inside his heart. I made the mistake of asking his pediatrician if it was okay that his lower lip sometimes looked a little blue. She left the room and returned a few minutes later, saying she had just called a friend who was a pediatric cardiologist and that he would see us right away, "just to be safe." My husband and I drove our newborn straight to his office. The doctor was all business as he clamped some miniature devices to our baby's fingertips. The blood oxygenation looked just fine, he assured us, adding, "as long as you're here, why don't we do an echocardiogram?" I was wobbly from childbirth and lack of sleep, so I didn't object. He held a device to my baby's bare chest and a video screen burst into life, showing throbbing swirls of red and blue. The colors indicated the direction of blood flow, the doctor explained, clearly proud of his sophisticated machine. Our son's heart had a small hole between two chambers, he said, quickly adding that this was unremarkable and fairly common. The hole would probably close—but just to be sure, he wanted to see us again in three months. We thanked him and left. In the parking lot, steadying myself against the car, I asked my husband, "What just happened? What *was* that? Did that doctor just do a totally unnecessary and expensive medical test and tell a couple of new parents that their precious bundle of joy has a *hole* in his heart?" My husband grimly said, "We are never going back there again."

We never did. And now my perfectly healthy son is six years old, old enough to start to learn about the heart. In the car, on the way to the grocery store, my husband tells him that we'll be getting special foods to drop off at the house of my son's close friend, as a present for this family because they're going through a tough time. My son says, tentatively, "There is a problem with the baby's heart." His statement of fact is a question. We did not know that he knew. Obviously, he was aware that his friend was expecting a baby brother, because he'd watched the mom's belly get bigger. What we hadn't realized is that our son found out about the fetus's heart defect before we did. His friend must have told him as they made paper airplanes or played with Legos. My husband says, "Yes, the baby's heart has a problem that can't be fixed, and he will die soon after he is born." My son abruptly begins to sob.

What he has started to learn about the heart, but can never learn, is that the heart beats and beats and beats and beats, never stopping, until it stops. When I was thirty years old, I got a phone call. It was a nurse at the hospital where my husband's mother was in the intensive care unit. She had cancer, and other medical problems, and fluid was building up in her lungs. Thankfully, her caregivers seemed generous with the morphine, and she was mostly unconscious. Now the nurse was telling me that they'd just given her a dose of morphine and her heart rate was dropping quickly. "I can't reach her husband or her son right now," I stammered, "They're out at a concert and don't have cell phones." Actually, the nurse told me, the heart had just stopped. "Oh," I said. I wasn't sure what I was supposed to do or say. I asked if they needed the hospital room right away. Maybe they could just disconnect all the tubes and wires and leave her in the bed, just for a little while, until I could send her family? The nurse said okay, they would do that, and we hung up. I waited alone in the house where my husband had been a child. Across town, with his father, he was listening to Brahms. Amidst the gorgeous music, inside his chest, was an imperceptible *pum pum pum*.

It is almost time for my children to sleep. We've turned off the overhead light and lie there together in the half-dark bedroom as I finish reading the book about the heart. I watch as my daughter follows its suggestion and tries to take her own pulse. Her fingers hunt for the right spot on her wrist. I can see she is pushing down too hard. My son touches his wrist more gently. Together we are quiet, attentive. "Are you finding it?" I whisper, moving my own hand to show them how it's done. I add, "Sometimes it's easier here, like this, in your neck." My son and daughter run their fingertips across their skin, pressing gently here and there, searching without success for the constant, elusive pulsing of their hearts.

Spider at the window

My first impressions of the spider: It was big and it was smart.

The brownish spider was nearly 2 inches long. It had constructed a thick, complicated web in the corner of my kitchen window's frame, outside of the house. On the wall inside, right next to the web, is a light fixture that almost always gets switched on at night.

"Look at this amazing spider!" I told my son, who was four years old. He dutifully peered out at the spider huddled against the glass. Then he went back to work on his own web-like creation, a "machine" made of strings that stretched across the room. He had tied the ends to the fridge and the oven and the sink. He was using more string to attach a broom and various pots and pans. His machines often make it difficult for the rest of the family to move through the house.

For weeks, every now and then, I checked on the spider.

At first, because of its size, I thought it might be a wolf spider. Then, on a website called *Spiderzrule*, I read that wolf spiders don't have webs and that they're often confused with funnel weaver spiders.

I now saw that part of the spider's web formed a funnel. It looked like a tiny frozen tornado. The spider spent its days at the bottom, which was right in the corner of the window. At night, the spider crawled up the funnel to the edge of a flat sheet of web. I learned that this sheet was not sticky. Instead, when an insect flew into its entangling filaments, the spider would rush out, bite its prey to inject venom, and drag the subdued victim back to the safety of the funnel.

I never saw that. Still, I admired the spider. I liked its black stripes, the little hairs on its legs, and its prominent spinnerets. On the far side of its web it had shed two translucent exoskeletons that looked like ghost spiders. I was pleased to read that funnel spiders could live for a year or more, always in the same web.

The spider's arrival coincided with a period of intense introspection. In the dark autumn mornings, the spider kept me company as I drank my coffee at the kitchen table while everyone else slept on, under warm covers upstairs. I would write a daily stream of consciousness on a legal pad and occasionally get up to pour more coffee and observe the spider. Sometimes it was curled up at the bottom of its funnel. Other times, it was at the top, poised to strike.

These mornings reminded me of what it had felt like to be pregnant and working at the office. I would feel the fetus lurching around inside the warm, pulsating universe that was my body, pursuing its own mysterious agenda and thinking its alien thoughts, while out in the bright world I typed away at a keyboard, writing about exploding galaxies and prehistoric fish and hidden mountains discovered under distant glaciers.

One night in November I gave a public lecture on science reporting. Outside the auditorium, someone had set up a hands-on science exhibit about spiders. The lady behind the table wore spider earrings and was showing off a tarantula. She briskly confirmed the identity of my spider—"if there's a funnel, it's a funnel weaver"—and told me that if it had a large abdomen, that meant the spider was female. My spider was a she!

Abruptly the spider lady said, not unkindly, "That spider's going to die tonight, you know."

The weather report predicted a freeze, she noted. "But I thought these spiders could live for a long time!" I protested. "Cheer up," she told me. "She'll have made an egg sac and you'll have babies to replace her. Didn't you ever read *Charlotte's Web*?"

I was insulted by the idea that my spider could be replaced. Yes, I had read *Charlotte's Web*. Who can forget Charlotte's wry, unsentimental acceptance of her fate: "After all, what's a life, anyway? We're born, we live a little while, we die." Recall how she died alone, after summoning her last strength to wave at Wilbur and whisper, "Good-bye."

I respected how E. B. White had kept her spidery; she relished killing and drinking blood. When the illustrator initially drew her with a woman's face, White rejected that idea and personally sketched in some dots and lines to make her look like a spider. But Charlotte wasn't just a spider. As the book put it: "It is not often that someone comes along who is a true friend and a good writer. Charlotte was both." My spider was neither. My spider, as far as I can tell, didn't know I existed.

As I went about my ephemeral chores—washing the dishes, hunting for plastic toy animals, wiping snotty noses, filing late-night copy—the spider existed nearby, motionless in her strange silken world.

If she paid attention at all as I put my face against the glass to gaze at her, less than an inch away, I am certain her weak visual system merely registered me as some random moving blobs that came and went and meant nothing.

She did not know me and I did not know her, even though I watched her and I loved her. I tried to imagine her perceptual world, with its thrumming language of vibrations and the taste of fly. I wondered about her inner experience, what she thought as she crouched in her funnel, whether she had dreams, whether she felt the increasing cold and understood.

My spider did not die that first freezing night. For several days she stayed in her funnel. I detected slight changes in position that meant she was alive. I dared to hope she was entering some kind of suspended animation. Perhaps she would survive the winter. Some spiders make their own antifreeze, I read.

I imagined that if she did die, I would go out and carefully remove her body from the funnel with a pair of tweezers. I would cup her dry, hardened body in the palm of my hand. I would use a magnifying glass to examine her and peer into her eight eyes. I wanted to get as close to her as possible.

Then one morning I went downstairs and checked the web,

and she was gone. This was unprecedented. I went outside and immediately spotted her on the other side of the window frame, a few inches down from another, much smaller spider. Neither of them moved.

That night she was back in her web, deep in her funnel, scrunched down smaller than I'd ever seen her. It was the coldest night since February. I was relieved she had returned. But the coming winter, plus her inexplicable foray across the window, left me feeling uneasy. I told my psychiatrist and he said, "Somehow I suspect that this spider story isn't going to have a happy ending." I replied, "Nature generally doesn't." Every week we sit in opposite corners of a quiet room and look at each other. We never touch, and I wonder what he thinks of me.

The next day the spider's web was again empty. Night came and she did not return.

In the morning, I checked the funnel, knowing that she wouldn't be there. I didn't think I could find her but I went outside into the cold anyway, shutting the door firmly behind me. Standing on the back porch, I studied the architecture of her web from this different perspective. I saw a bit of blue iridescence, some leftover piece of an insect's thorax. Deeper in the threads I saw a velvety dead bee.

I searched all around the window frame, inspecting every crevice. Way down at the bottom, far from her web, I saw a crack filled with about a dozen balls made of silk. Each was about the size of a pea. I didn't believe that these belonged to my spider, but I crouched down to study them.

Suddenly I sensed a dark blur behind the glass and I looked up through the window to see my husband's kind, familiar face looming just beyond the spider's web. He smiled down at me and raised his eyebrows as if to ask, "Well?" I shook my head. He made a face of mock dismay that softened into fond concern. He gazed at me for a moment, then turned away from the glass and disappeared.

A life so precarious as a flea's

> To produce a mighty book, you must choose a mighty theme. No great and enduring volume can ever be written on the flea, though many there be who have tried it.
>
> —Herman Melville, *Moby-Dick*

The one book I failed to read in my high school American literature class was *Moby-Dick*. I dutifully absorbed *The Great Gatsby* and *A Streetcar Named Desire* and *Manhattan Transfer*, but Melville mystified me. I resorted to CliffsNotes to get the basic outline of the plot and characters, memorizing just enough to pass the test. Decades later, I picked up the book again and found that Melville's gorgeous, Shakespearean sentences now dazzled; his metaphorical sleights of hand made me dizzy; the briny profundity in once-tedious nautical passages made me want to weep. I reread the book every summer during my family's annual beach vacation, and I'll look up from, say, Chapter 87, *The Grand Armada*, with its evocative scene of mothers and young whales drifting peacefully together, surrounded by swirling undercurrents of impending doom, to squint against the sun's glare and see my own children squealing and splashing in the waves. Yet despite the affection I feel for this bizarre book and all its wisdom and beauty, I always wince when I get to the part where Melville dismisses, so casually, the flea.

In 1850, when Melville spent each day writing *Moby-Dick* at his beloved Arrowhead farmhouse in Massachusetts, in the company of his pregnant wife and infant child and aging mother and two unmarried sisters, it is perhaps understandable that fleas did seem puny, even comical. Melville didn't live to see scientists uncover the flea's true nature, although, I would argue, enough hints were available that he should have known better. To me, it is obvious why a "great and enduring volume" could be written on the flea (despite Melville's remark that "many there be who have tried it," *did* anyone?). But my understanding of the flea developed slowly, without me even recognizing the source of its itch, just like it took me half a human life span to apprehend the nameless horror of the white whale.

Fleas were largely theoretical beasts when I was a girl. They were tiny and could jump and their bites were itchy; I knew all that in the same way that I knew unicorns were magical white horses with spiral horns that could only be tamed by a virgin. As a kid, I'd seen cartoons that showed dogs scratching mightily, the comic shorthand for fleas. But I'd never seen a real, live flea, even though I loved animals of all kinds and had an enormous tabby cat named Tiger. Tiger roamed in the woods behind our house on a suburban cul-de-sac. At night, he'd jump onto the railing of our back porch, then onto the air conditioner box jutting out from the transom over the back door, then up onto the roof of our garage. From there, he could saunter over to my bedroom window, which looked out over the roof and was right beside my bed. I'd be lying there reading, surrounded by stuffed animals, and hear him meow. I'd push up the window and he'd step across the sill and onto the quilt, rubbing against me and purring. I'd pet and kiss him tenderly, oblivious to any remnant of murdered mole on his paws or whiskered face, as hidden fleas no doubt crawled through his gray striped fur.

He'd disappear for days, and I'd worry. Talking about this with my parents one night, sitting in the back seat of the family Buick, I said that Tiger would go off without warning, then casually show up outside my bedroom and act like he'd never been gone.

"Well, get used to it, because that's how it'll be with a lot of men," said my father in a wry tone as he steered the car into our driveway. My mother immediately shushed him, but also laughed. Her half-serious indignation caught my attention; like all children, I was highly attuned to whatever adults tried to hide.

My parents never expressed any concern about fleas, however, either out loud or sotto voce, so I assumed the rubbery flea and tick collar that my cat wore kept them away, like garlic warding off vampires. I should have reflected on the fact that a new collar's horrible smell always dissipated rather quickly, and the collar seemed to have little effect on the ticks. Occasionally I'd feel a tick's distended body, hard and engorged with blood, beneath my cat's silky fur. Ticks existed in a way that fleas did not.

As I got older, I read about fleas in books and magazines like *Ranger Rick*, which mostly stressed fleas' incredible ability to jump farther than fifty times their own body length, and showed close-up photos of fleas looking like flattened armadillos, with overlapping protective plates and thick, thornlike hairs protruding here and there. These color photos were just as mind-bending for me as Robert Hooke's detailed black-and-white drawing of a flea must have been for readers in 1665 who picked up his famous book *Micrographia: Some Physiological Descriptions of Minute Bodies Made by Magnifying Glasses with Observations and Inquiries Thereupon*.

In this book, Hooke described how the tip of a needle and the edge of a razor looked under a microscope, and he was the first to use the word *cell* to describe the building blocks of tissue (because the boxlike shapes he saw in a thin section of cork reminded him of cellulae, the rooms that monks lived in). The highlight of his many close-up illustrations was a large, foldout page filled with an elaborate, larger-than-life flea. In Hooke's time, flea bites were a mostly unremarkable part of everyday life; the nightly "flea hunt" became a popular theme of painters, because it allowed them to depict beautiful, half-naked women in their boudoirs, searching their pale white skin for vermin. But no one had seen anything like Hooke's illustration, which revealed the humble flea to be a surprisingly burly and bristled beast. "The strength and beauty of this small creature, had it no other relation at all to man, would deserve a description," wrote Hooke. "The *Microscope* manifests it to be all over adorn'd with a curiously polish'd suit of sable Armour, neatly joined, and beset with multitudes of sharp pins, shap'd almost like Porcupine's Quills." This caused a sensation. Diarist Samuel Pepys stayed up late one night reading *Micrographia* and called it "the most ingenious book that I ever read in my life."

No one realized, then, that fleas carried the Black Death, the terrifying illness that periodically swept through Europe. It was about to culminate in the Great Plague of 1665 that killed a quarter of the people living in London, just months after the publication of Hooke's bestseller. The flea as a bringer of mass destruction was not recognized for another 230 years, after science had firmly established that germs cause disease.

The first clue that fleas might be involved in plague came in 1897, when a Japanese doctor named Masanori Ogata remarked, "one should pay attention to insects like fleas for, as the rat becomes cold after death, they leave their host and may transmit the plague virus directly to man." A French physician, Paul-Louis Simond, had been thinking along the same lines. While in Karachi during an outbreak, Simond did experiments with live animals, writing, "I was fortunate enough to catch a plague-infected rat in the home of a plague victim. In the rat's fur there were several fleas running around. I took advantage of the generosity of a cat I found stalking the hotel premises, borrowing some fleas from it." After confining the sick rat in a specially designed apparatus and throwing in the cat's fleas to ensure it was well covered in parasites, he waited. The sick rat died, and fleas from it traveled through the device over to a healthy rat that subsequently began showing signs of the plague. "That day, 2 June 1898," wrote Simond, "I felt an emotion that was inexpressible in the face of the thought that I had uncovered a secret that had tormented man since the appearance of plague in the world."

And what a secret! Until then, like Melville, most people saw fleas as inconsequential, even humorous. In the mid-nineteenth century, as Melville scribbled out *Moby-Dick* and handed pages to his wife and sisters, who slaved away as his scriveners and cooked his meals and took care of his every need, showmen like Louis Bertolotto amused crowds in London and New York with flea circuses. Fleas collared with thin metal wires pulled chariots or walked on tightropes or even seemed to juggle balls. In reality, these poor fleas were just scrabbling their legs wildly because they were trying to escape, but Bertolotto claimed to "educate," or train the fleas, a claim that a few scientists of the day were quick to dismiss, even though Bertolotto had more experience with fleas than most naturalists ever would. Charles Dickens, after visiting Bertolotto's shop in 1856, wrote that "sometimes he has nearly all his fleas on the backs of his hands at the same moment, all biting and sucking away."

Like Melville, Bertolotto surrounded himself with females. "Supporters of the women's rights movement will be delighted to know, that my performing troupe all consists of females, as I have found the males utterly worthless," Bertolotto once wrote, adding, "some people have raised a cry, that I am cruel to my fleas, in making them do as they do; but this idea exists, only in their imagination, for few masters give so freely of their blood." The public that loved his fleas' antics proved less welcoming to *Moby-Dick*. The novel bombed, and Melville became so mentally unstable and abusive—whether he once drunkenly pushed his wife down some stairs is controversial, but the fact that it's even debated is saying something—that his wife's brother and a minister conspired to get her away by kidnapping her, so she could not be charged with desertion. In the end, Lizzie Melville decided to stay with her husband, either out of love or loyalty or practicality or some complicated mix of emotions, and she "kept her husband's literary legacy alive," according to Melville biographer Laurie Robertson-Lorant, who described the marriage as long and complex: "She was as heroic in her way as he was in his, and in the end she understood him best and took whatever secrets she had with her to the grave. Carved into the back panel of her writing desk are these wise and poignant words, 'To know all is to forgive all.' "

When *Moby-Dick* began to be recognized as a work of American genius in the 1920s, flea circuses were still going strong. One history of them notes that "a select few featured both dead whale and live flea, contrasting the monstrous with the minuscule." In 1935, for example, the Eureka Whaling Company put a 55-foot, 68-ton whale "in a state of perfect preservation" on a specially built train car. As this marine marvel traveled across the United States, it was accompanied by Madame Sirwell's European Flea Circus. One town's newspaper offered a coupon for free admission to any child under the age of twelve. By the time I was that age, in 1986, flea circuses had become a thing of the past. The comedy of performing fleas lingered on, however, and I knew of them from shows like *Sesame Street*, which ran a cartoon called "F is for Flea Circus." I had no idea that fleas had ever been put in chains and forced to perform. From what I'd been able to glean, flea circuses were scams, and I wasn't completely wrong; comedians and magicians had contrived "humbug" flea circuses set up in suitcases that used magnets or mechanical contraptions to make it look as though, say, a flea named Colossus was pushing a ball around a miniature circus ring. In order for this type of "flea" act to have any possible chance of fooling an audience, my childhood-self assumed, fleas must be vanishingly small—almost imperceptible to the naked eye. No wonder I had never seen one.

On Sundays, my hometown's volunteer fire department filled its parking lot with a flea market—to this day, there's a yellow sign with an arrow tacked up by the squat brick firehouse that says "FLEA ENTRANCE." As a kid, I'd often ask my parents to drop me off there after church, a request that invariably involved jokes about buying fleas. I'd walk among the rough wooden tables covered in shabby antiques and cheap housewares, looking for treasures. Instead of the nonexistent fleas, I brought home porcelain cats. My favorite was a white Persian cat with blue eyes. This cat sat on the dresser in my bedroom until one day when it broke into pieces. I don't remember how; what I do remember is my mother coming out of her study, holding the resurrected white cat in her hands, and my astonishment at her ability to reassemble the jagged pieces. (This china cat now lives in my dining room; its stoic, dark-lined eyes stare out at me from the top of a cabinet. I see that my mother used something like Elmer's Glue and her fix-it job was rather sloppy, with drips down the side that solidified and have yellowed with age.)

Other precious items proved less amenable to repair.

Late one night, I'd been in my bedroom, making some kind of witchy potion out of ingredients that included cinnamon mouthwash, when I'd accidentally spilled this concoction onto a white rabbit fur. It was the kind of pelt you could buy for a few dollars as a souvenir at a faux general store in Colonial Williamsburg, but to me it seemed incredibly valuable; it was *real fur*. As the red stain spread over its luxurious softness, I panicked. I crept into my parents' room and woke up my father, who was snoring away in their huge four-poster bed. He got up immediately, without complaint. He stood there in the bathroom in his pajamas, a half-asleep, middle-aged man blinking against the light, silently rinsing the rabbit pelt beneath the tap and scrubbing it gently with a bar of soap. I watched him ruffle the fur with his fingers as it lightened from dark red to a pale pink, a stubborn stain that was almost back to white but not quite.

White fur, it turns out, is what I needed to finally see fleas. Our pet cats were always tabbies, or calico, or ginger, or black. At some point, though, I found myself caressing a cat with white fur, maybe a stray. I noticed teeny black specks. Were these fleas? Fascinated, I looked closer, parting the fur to investigate. I saw more dark flecks, but they didn't move. I kept running my fingers through the whiteness, checking here and there, when I saw a much larger dark shape squirm away. *Was this a flea?* To my eyes, it seemed enormous. Ignoring the protestations of the cat, which objected to my prodding and hair pulling, I single-mindedly pursued the flea as its flattened body, the color of dried blood, slid through the white fur. In this, its natural element, it moved swiftly; "its keel-shaped forehead forcing its way through the undergrowth like the bows of a boat cleaving the waves," as one famous entomologist once put it. I parted the soft fluff again and again until I managed to seize the flea between my pointer and thumb, sliding the creature up along a hair until it was off and imprisoned within my pinch. Then, opening my fingers slightly, I peeked in. The flea immediately leapt away, and I, entranced, realized that I had witnessed the flea's legendary jump. I hunted through the poor cat's fur some more. I felt no revulsion. I so much wanted to see a flea. I could only catch quick glimpses, either when its small brownish-black body was exposed on the cat's skin before it wriggled deeper into the white or when it perched on my thumb briefly before springing away.

That special talent of fleas—jumping—was, strangely, almost totally unexplored in the first book on fleas written in English. *The Flea*, published by Harold Russell in 1913, was a slim volume, in part because zoologists had long neglected the lowly flea and had collected so little information. "The statements about these insects in the general text-books of entomology are frequently antiquated and inaccurate," wrote Russell. The recently discovered link to plague, however, had given the flea a newfound gravitas, and interest among scientists had started to grow. Russell went on at length about the mouthparts of the flea, which let it pierce the skin and suck blood, and discussed, in great detail, how different species of birds and mammals and even reptiles all host different species of fleas that sometimes end up on unexpected animals. ("The preference which fleas show for certain animals, and the repulsion which they manifest on being allowed to suck blood from an unaccustomed host, lead one to believe that they have a sense of taste.") When it came to their jumping, however, Russell merely remarked that "it has often been pointed out that if men had the leaping powers of some fleas they would bound with ease backwards and forwards over the cross on the top of St. Paul's Cathedral." He barely described the flea's six legs, except to note that the hind ones were the largest, concluding lamely, "They are the organs of hopping."

A half century later, that state of scientific ignorance was not at all acceptable to Miriam Rothschild, perhaps the only flea researcher to ever become famous for an investigation of this creature—and, notably, a woman working in science when that was a rarity. An eccentric heiress in the wealthy Rothschild family, she was also a throwback to an earlier time, when self-educated naturalists could use their own fortunes and talents to pursue eclectic research for kicks. Her banker father, Charles Rothschild, had amassed what had to be the most important collection of fleas in the world, and he had described and named the flea that is a primary carrier of the bubonic plague. His flea collection included not only hundreds of species and subspecies, but also a small number of kitschy "dressed fleas" from Mexico, where folk artists adorned fleas with tiny clothes; as one museum noted, the most popular forms of this art were "bride and groom or farmer and wife sets."

Charles Rothschild kept his daughter Miriam out of school but encouraged her interest in living creatures, treating her as an adult when it came to collecting and studying parasites. After he died by suicide when she was fifteen, she briefly considered becoming a writer. Instead, after dissecting a frog with her brother, she decided to pursue science, and put together a multivolume catalog describing her father's flea collection that took several decades to complete. Asked why she spent so much time on fleas, she told an interviewer, "That was purely a filial devotion. My father was a flea man and I became a flea girl, it's as simple as that."

She became known as the "Queen of the Fleas," admired by the public for being unimaginably rich and yet also being down-to-earth enough to keep fleas in plastic bags in her own bedroom, so that she could watch their activities without having to worry that her six children would annoy them. (Her marriage lasted fourteen years. When once asked if she had gotten hitched just to have a "stud," she replied,"Good Heavens no! It was a love affair, a real love affair.") She studied rabbit fleas to try to understand why no one had been able to breed this particular species, keeping the fleas alive by letting them feed on white rabbits—finding, as I did, that fleas are much easier to spot against the white.

Perhaps her own experiences with pregnancy clued her in to its potential importance in the rabbit and flea relationship; she realized that a female flea had to suck blood from a pregnant rabbit in order to get a hormone it needed for its eggs to mature. This was the first known example of a parasite's reproduction being dependent on that of its host. In the 1960s, Rothschild zeroed in on the flea's jump, taking the first high-speed photos of jumping fleas and minutely dissecting the flea's anatomy. She authored hundreds of scientific papers, intrepidly burrowing into the male-dominated field of entomology and serving as the first female president of the Royal Entomological Society, all while promoting conservation and wildflower gardens to support beneficial insects. She moved through these gardens "like a ship in full sail, dressed in her favourite purple or sea-green gowns," one obituary recounted, adding, "a lifelong atheist, she admitted that she had been tempted to believe in a creator when she discovered that the flea had a penis."

Now, Melville thought it was impressive that a sperm whale had a phallus "longer than a Kentuckian is tall," as he put it in *Moby-Dick*, in a bawdy chapter called "The Cassock" that seems to mock the Christian church with its depiction of a sacred robe being fashioned from the skin of a whale's penis. But the whale's "grandissimus" seems rather humdrum when it's considered in proportion to the size of the rest of the whale. The body of a male flea, in contrast, contains a curled-up penis that, when stretched out, can be more than two and a half times the length of its body. Rothschild, describing the flea's penis, called it "a structure of extraordinary complexity—in fact, it is the most complex genital organ to be found in any insect. The more one considers it, it is difficult to understand how such a structure could have been evolved." After watching copulating fleas for hours, she found herself agreeing with an American morphologist who had done the same and concluded, "Truly, the thing does not make sense."

Despite her professed admiration for the male flea's equipment, when she published *A Colour Atlas of Insect Tissues via the Flea* in 1984, Rothschild chose as its cover art not a penis but rather a multicolored image of a pregnant flea's oviduct and vagina. "It's the only book in the world which has the vagina of a flea on the cover," she once said proudly, "but you must say it's beautiful." As a fan of fleas, with a keen interest in their sex life, Rothschild would have been aware of John Donne's poem "The Flea," written in the 1590s, in which a man trying to seduce a virgin points to a flea that has bitten them both and mingled their blood:

This flea is you and I, and this
Our marriage bed, and marriage temple is;
Though parents grudge, and you, w'are met,
And cloistered in these living walls of jet.

The flea might seem an unlikely source of eroticism, but Donne's verses were part of a long history of literary soft porn involving fleas, whose ability to freely travel under women's clothes while sucking on their skin and swelling with blood would probably have inspired innuendo even if the French word for flea, *puce*, didn't sound so similar to the word for maiden, *pucelle*, or the word for maidenhead, *pucelage*, and deflowering, *depuceler*. (Elsewhere in his writings, Donne used fleas for a decidedly less amorous comparison, saying that women are like "fleas sucking our very blood, who leave not our most retired places free from their familiarity, yet for all their fellowship will they never be tamed or commanded by us.") But the connection between fleas and sex goes back even further, before human imagination. Recent DNA analysis has revealed that the closest relative of the flea is the scorpion fly, which means that the ancestor of modern fleas likely started out sucking nectar from the sex organs of plants. Then, between 290 and 165 million years ago, for reasons unknown, proto-fleas lost their taste for simple sweetness and switched to blood. Fossils of nearly-1-inch-long primitive fleas have been found from the Jurassic period, suggesting that they might have feasted on some of the largest animals to have lived on Earth.

The largest extant flea lives in the Pacific Northwest, where it reaches nearly half an inch in length and specializes on a rodent called the mountain beaver. Not too long ago, a biologist named Merrill Peterson became obsessed with getting the first ever photograph of a living specimen of this flea. "And that is how I, an asthmatic writer, ended up with my lips on a flea-collecting device powered by sharp inhalation, watching, terrified, while Merrill, a man deeply averse to touching most mammals, wrangled a toothy, clawing wild mountain beaver inside of a basmati rice bag," his wife, a science writer named Carol Kaesuk Yoon, wrote in the *New York Times*. Her husband struggled to keep the trapped rodent's head in the sack while he combed through the hair on its back end. "Merrill asked me to cut away as much of the bag as possible so he could comb more of the mountain beaver. And though burlap was all that stood between us and teeth and claw, I cut," wrote Yoon, a paragon of marital devotion. "He combed the underside of the beast. Something brown tumbled down. That's when the love of my life—normally eloquent and articulate—began shouting, 'Flea! Big flea!' " She gamely sucked at the collection device, much like a flea going after blood, as her husband urged her to suck harder. She finally managed to inhale a humongous flea into the device's chamber, thrilling her spouse even as she herself veered "between the urge to giggle hysterically and the urge to vomit."

My early flea experiences didn't involve wild animals, except for the fleas themselves; perhaps this is why I hadn't felt any sense of threat. That first flea I had spotted and tracked in white fur clearly had no interest in me; I had been the one hunting it, while its streamlined body dove ever deeper. Like Moby Dick, this animal wanted to escape, nothing more. I had intuited, as a child, what science did not establish until I was in high school. In his opus *The Flea*, Harold Russell had asserted that fleas "pass but a portion of their lives on their hosts and frequently take occasion to hop on and off." But in 1989, an entomologist at Kansas State University named Michael Dryden did studies with cat fleas showing that fleas have zero desire to hop on and off. Any flea fortunate enough to land on a reliable source of blood meals will never willingly leave; an adult flea's sole goal is to go about its business on an animal's skin for the two to three months it has before dying. That business, put simply, is to mate and to create; the female *to-do* list consists of eating about fifteen times her own body weight in blood every day and occasionally pushing out a pearl-white egg.

Since flea eggs have no way to attach to the host animal, they fall onto the ground or floor or bedding, as does the dried skin and feces of the parents, which serve as critical food when the white, wormlike larvae hatch. Eating their parents' waste turns out to be essential for young fleas, as it contains unmetabolized blood with iron that the larvae can use to build a safe shelter where they can grow and change. The young fleas spin a sticky cocoon that starts out white but quickly gets dirtied and camouflaged by dust and debris. Inside this protective cover, the immature larva transforms into an adult. A mature flea can wait inside its cocoon without food for as much as a year, until it senses vibrations and an increase in carbon dioxide that might indicate the presence of a blood-filled animal that could make a suitable home. These signals, unfortunately, are ambiguous; an emerging adult flea has to take a great hopeful leap, flinging itself toward an uncertain future, because after it leaves its place of safety, a flea needs to start feeding within hours in order to survive. As one scientist observed, "It is more than likely that, in a life so precarious as a flea's, speculative jumping plays a very large part."

If I had understood, as a young adult, that flea eggs fell off infested animals like salt from a shaker, I might have felt some trepidation about letting strange cats climb all over me. Since I did not know, I had no qualms about cuddling with random cats or taking in strays, just as I had no qualms about getting drunk and landing in bed with this man or that. In college, a pathetic tabby kitten followed me down the alley behind my apartment building, so I brought it inside. The kitten slept with me and scampered around the living room, playing with string. I reluctantly brought it to an animal shelter when I couldn't find anyone to take it in permanently before I had to leave for the summer and go back to my hometown. A few weeks after that, my roommate called and said that a flea infestation in our apartment had forced her to hire an exterminator. She told me that fleas had been leaping out from between the wooden floorboards and biting her feet and ankles. I apologized and wrote a check to reimburse her. Secretly, though, I thought she was a bit of a neat freak who'd had some kind of paranoid overreaction. I'd lived with cats my entire life and never had experienced anything like that. I'd never even *heard* of anyone calling an exterminator for fleas.

I assumed that she'd seen a flea or two and panicked, or possibly just imagined their existence, in the way that thinking about fleas long enough can create an urge to scratch. Without the instantly accessible information that exists on the Internet today, I couldn't easily look up facts about the flea life cycle. I didn't know that the famous entomologist Karl von Frisch, who discovered that bees communicate by waggle-dancing, had once described sharing a hotel room with a colleague and finding the room to be "alive with fleas." Each night the two scientists would walk around in their bare feet and nightshirts, letting themselves be attacked, and then count up their haul. "My 'bag' every night amounted to a mere four or five, whereas my friend always caught thirty or forty," Frisch wrote. "They seemed to prefer him." When I came back to my university apartment for the fall semester, the place was flea-free, but some of my clothes were packed inside plastic bags, sealed up by the exterminator. I eyed them warily, eventually putting them in the freezer for a while.

And I kept taking in strays. Visiting the man I later would marry, I befriended a thin white and black cat outside his rowhouse. When I sat on the front steps, the cat curled up in my lap, ecstatically purring. I told my soon-to-be-husband, "I think I have a cat now." Together, we drove the cat back to my place in Baltimore. Sleeping at night with my lover curled up on one side of me and the cat curled up on the other, I felt at peace. Over the years, my husband and I adopted other cats together, and despite their inescapable medical complications and feline neuroses, our home was never tormented by fleas. Maybe it was the new-fangled treatments, designed to be dripped onto the back of a cat's neck so that the chemicals could then creep across the skin (making us wonder if a wayward drop might slide along *our* skin). Or maybe we were just lucky. Whenever fleas came up in conversation with our two children—who, like me, were growing up with no firsthand knowledge of fleas—my husband would recite a bit of doggerel that had lodged in his head, waiting for the right opportunity to jump out:

Great fleas have little fleas upon their backs to bite 'em,
And little fleas have lesser fleas, and so ad infinitum.

This poem dates back to 1872, to a book called the *Budget of Paradoxes*, by mathematician Augustus de Morgan. My husband usually left off the rest of it:

And the great fleas themselves, in turn, have greater fleas to go on;
While these again have greater still, and greater still, and so on.

These verses made up part of a chapter entitled "Are Atoms Worlds?" and were used to illustrate the notion that, just as atoms are made of even smaller particles, and smaller particles, and on and on, perhaps our planets and stars are "particles in some larger universe, and so up, forever."

Morgan's verses about fleas and the nature of the universe, large and small, echo the words of an earlier poem published in 1733 by Jonathan Swift, which included the lines:

So, naturalists observe, a flea
Has smaller fleas that on him prey;

And these have smaller still to bite 'em,
And so proceed ad infinitum.

But Swift wasn't writing about fleas or atoms or cosmology or science. He was writing about writing and the lamentable tendency of great poets to be criticized or second-guessed by lesser talents:

Thus every poet, in his kind,
Is bit by him that comes behind:
Who, though too little to be seen,
Can teaze, and gall, and give the spleen.

This state of affairs in the world of letters was far different from the ordinary state of nature, Swift noted, where large animals such as "a whale of moderate size" could simply swallow up little fish whole.

And so it is that a flea-like writer like me can bite at a genius like Melville, taking a book as rich and innovative and enduring as *Moby-Dick* and choosing to argue over the significance of one brief phrase: "no great and enduring volume can ever be written on the flea." But like a flea, the smallness of this phrase is misleading; the dismissal of the flea is the final thought in a key passage that seems to express Melville's ultimate literary purpose in writing about the white whale:

One often hears of writers that rise and swell with their subject, though it may seem but an ordinary one. How, then, with me, writing of this Leviathan? Unconsciously my chirography expands into placard capitals. Give me a condor's quill! Give me Vesuvius' crater for an inkstand! Friends, hold my arms! For in the mere act of penning my thoughts of this Leviathan, they weary me, and make me faint with their outreaching comprehensiveness of sweep, as if to include the whole circle of the sciences, and all the generations of whales, and men, and mastodons, past, present, and to come, with all the revolving panoramas of empire on earth, and throughout the whole universe, not excluding its suburbs. Such, and so magnifying, is the virtue of a large and liberal theme! We expand to its bulk.

Maybe it's a peculiarly male preoccupation, this obsession with swelling and rising and expanding. Is there not something to be said for the universe and all generations re-created in the small and unremarkable? Is it, perhaps, peculiarly female of me to focus on the potent symbolic power of less exalted lives that might exist, waiting and unseen, in the thin cracks between the floorboards of an ordinary home? Is it unfair to point out that even if fleas got little respect during Melville's lifetime, some thinkers had nonetheless managed to look at this insect and see everything from love to sex to marriage to death and the mind-bending reality of infinity?

Take the idea that fleas have their own parasites, and so on, *ad infinitum*. That originally came from Antonie van Leeuwenhock, a microscopy pioneer who recounted seeing a tiny mite on flea larvae. This was just one detail in an amazing letter that Leeuwenhoek wrote to the fellows of London's Royal Society in 1693, describing every step of flea reproduction after imprisoning male and female fleas within a glass tube to watch their coming together. "The movements which the male made for about half an hour at a stretch were as indefatigable and as lustful as one might see in any beast whatsoever," Leeuwenhoek wrote. He dwelled on flea sex and eggs and larvae in part to discount the persistent belief, held through the centuries since Aristotle, that fleas and other insects spontaneously generated themselves from nonliving matter like sand or peat or dung. "Are we then to be burdened still with those old tales?" he asked the Royal Society fellows, clearly exasperated. "I entirely deny that Fleas could be formed from dust and piss, and I assert that in my view this is impossible." The flea was such a despised, trivial being, and yet, Leeuwenhoek maintained, "we see such a neat arrangement and perfection with which the Flea is endowed as exists in any large animal." (Even one as large as a whale.) The flea's minute perfection could be seen as even *more* impressive, even more fearfully made. Leeuwenhoek noted that a draftsman he had employed to draw the flea's leg or other parts "often burst out with the words: DEAR GOD, WHAT WONDERS THERE ARE IN SUCH A SMALL CREATURE!"

And alright, while it's true that pre-*Moby-Dick* science remained blissfully unaware that fleas had carried deadly plagues that had killed untold millions, how could the bloodsucking nature of this tiny vampire not convey something slightly sinister, not so easily written off as a joke? The poet William Blake felt it when he believed that he had been visited by the "ghost of a flea" during a seance in 1819. Fleas, the ghost told him, were inhabited by the souls of men who were "by nature bloodthirsty to excess." In a painting he made of this supernatural vision, Blake depicted a tiny, realistic flea between the feet of its demonic, humanoid, spirit form—a dark, naked, muscular body, with what looks like long braided pigtails, like a sailor's, stretching out its tongue to lap blood from a bowl.

What the white whale was to Melville has been hinted; what, at times, the flea was to me as yet remains unsaid. When I reached middle age, I'd sometimes bring my children to visit my parents at their farm, a big old place out in the country with a brick house and a garden surrounded by cow pastures. My husband and I had gotten married there, looking like two well-dressed folk art fleas, with me in my mother's white gown, being walked down the aisle by my father, who wore the same black tuxedo he had worn at his own wedding. (I was only twenty-six when I vowed to be faithful "til death do us part." It seems like a great speculative leap for one so young, but I must have felt some inchoate certainty that I'd need a loyal companion on my life's voyage, one willing to support me through any possibly ill-advised hunt for elusive or dangerous quarry. Even Melville began *Moby-Dick* with a marriage, albeit of the idolatrous, heathen kind.) My parents went to the farm less and less often as they got older and frailer. My father, during one late-night episode at the emergency room, asked me to help him arrange to get my mother a cat; specifically, a white Maine coon cat, which is what my mother had been longing for but never got because she thought they traveled too much to keep a pet. I soon found myself researching catteries and putting down a deposit on a very expensive kitten for my mother, whose acceptance of our plan seemed like a tacit acknowledgment that their carefree traveling days might be coming to an end. "We've never had a white cat," she told me, as if I didn't know, and I wondered if I was buying her the last cat she'd ever have, if I was trying in some white-cat way to put broken pieces together again, as she had once done for me.

One year, my parents hadn't been well enough to go to the farm at all, and when my husband and I arrived there with the kids in late summer, the house looked abandoned. The grass in the yard and fields had gone to seed; it was brown and taller than waist high; weeds strangled the garden's flowers. My husband went to see if he could get the lawn mower going and the kids ran off to watch some cows, while I went inside the house to scrounge up a snack. In the pantry, where my mother kept the same kind of cheese-and-peanut-butter crackers I'd taken to school for lunch as a kid, I found empty, gnawed-on cracker boxes covered with rodent droppings. I knew I should clean up the mess, but I found myself shutting the cabinet. I went to the bathroom. Sitting there on the toilet with my shorts pulled down, staring absent-mindedly at the floor, I felt something down by my ankle. Some kind of insect was biting me. I swatted it away. Then I felt it again, and, at the same time, on the other leg too.

Looking down, I saw black specks on my skin. I slapped at them; but they reappeared just as quickly as they left. A word floated up: fleas. That seemed impossible. No pets had ever lived at this farmhouse. The closest thing was a fake cat that my father had bought for my mother years before; it was curled up on a chair just outside the bathroom door. This imitation cat had a pink bow and closed eyes and was made of some hard body, like cardboard or plastic, covered in white fur. My daughter, who liked to stroke the fur despite the cat's apparent rigor mortis, asked me once, "Is this fur from a real cat?" and I assured her that no, it was real fur but probably from a rabbit—as if that made any difference. Standing up and pulling up my pants, while swatting at my legs, I looked around the bathroom, trying to understand where these mystery bugs came from. A pipe went from the back underside of the toilet into a surprisingly large and jagged opening in the wall's tiles. This hole led to darkness and the smell of something dank. Crouching down beside the toilet, staring into this hole, I remembered the rodent droppings. I pictured fleas squirming through greasy fur and jumping off rotting corpses, hidden behind the wall. At long last, I felt revulsion, as the nearness of my body seemed to encourage the fleas. More and more leapt toward me; I could feel them crawling on my skin and under my clothes. They began to bite. Appalled, I stood up and backed out of the bathroom, then turned and ran out of the house. Standing on the porch, leaning out over its wooden railing, I called and called for my husband. I know he would have come, but he couldn't hear me. He was too far away, somewhere out there in an ocean of tall dry grass that rustled and waved in the wind.

Everybody does it!

I used to draw geometric patterns compulsively. I have notebooks filled with hundreds, if not thousands, of intricate patterns, usually a rectangle or square divided up into symmetrical sides filled with alternating dark and white shapes. Mostly I made these at the office, while talking on the phone or sitting in a meeting. I'd make a square and then divide it into four equal triangles by drawing diagonal lines from each corner to the opposite one. I'd keep adding intersecting lines, one after another, creating an ever more complicated web.

Here's a scenario that repeated over and over: Someone who didn't know me well would stop by my desk, see a half-finished design sitting next to my keyboard, and say something like, "Wow, that's cool!" I'd then open a notebook to reveal that it was just one of many, many similar doodles. Flipping through the pages produced a blizzard of these perverse black-and-white snowflakes, and the person's expression would invariably change from intrigued surprise to something more like bewilderment or concern at the sheer number of hours I must have spent inking in tiny triangles and squares.

I usually doodled on notebook paper, using the side of an extra pen as a straight edge. Or, if I wanted to make circles, I'd feel around in my backpack for a coin to trace. I didn't care about the finished product. Once, an airline stewardess accidentally spilled some water on a sheet of paper that I had out on the tray table. She apologized profusely, perhaps imagining she'd ruined some sacred mandala. I assured her that after the flight I'd been planning to throw it in the garbage along with my empty mini pretzel bag. Another time, at a National Institutes of Health conference, an audience member sitting behind me had watched me doodling in my notebook during the presentations. During a break, he offered to buy my doodle. He said it reminded him of work he did on DNA arrays. I think that I ripped it out and handed it over for free. I craved the semi-hypnotic state I entered when making geometric patterns; afterward, looking at all the ink that marked up the page, I felt nothing.

I had such a reputation as a doodler at work that my colleagues immediately thought of me when the science editor wanted a report on a study of doodling that purported to show that this activity enhanced people's ability to remember information from a long, boring telephone message. (At least one study since then has challenged that idea, with a similar experimental design that found free-form doodling was actually associated with memory costs.) For a website video to accompany the NPR report, a colleague who was a videographer wanted to film me doodling. I went along with it, even though

I found the attention to this particular personal habit disconcerting, as if someone wanted to interview me about biting my nails. A random person who saw the online video wrote to ask if he could have some of these geometric designs, since I'd said that they meant nothing to me. I sent him a small pile, hoping he wouldn't turn out to be a stalker; he wrote back to say his kid had been excited to get them, which felt reassuring.

People's interest in my geometric doodles eventually made me wonder if I should try to "do something" with them—especially after I went to a museum in New York City and saw an exhibit that featured drawings eerily like my own. The American Folk Art Museum held an event called *Obsessive Drawing* that included five self-taught artists, all men, who filled pages with circles, dots, and geometric patterns, in order to, as the museum put it, "help them cope with illness, loss, loneliness, fear, and regret." Eugene Andolsek, for example, worked as an anxiety-prone stenographer for a railroad line by day. At night, for about fifty years, he sat at his kitchen table and used graph paper as a basis for making ever-more-complicated kaleidoscopic drawings. "The pictures were never displayed on his walls nor exhibited," noted one write-up about his work. "Once completed, the pictures held no interest for Andolsek and were put in the closet or a trunk. In fact, Andolsek did not think of himself as an artist nor saw any value in what he created beyond the desire to draw them each evening."

After going to that museum exhibit, I started doodling on plain white paper, to make the images more starkly black and white than they'd been when I did them on notebook paper lined with blue and pink. I'd take a piece of paper from the office printer, fold it into quarters, and then rip carefully along the folds. During idle moments in the workday, I'd fill in each rectangle. The inked scraps of papers started to pile up. I stored them in Tupperware. I thought that maybe I could try to use them to make actual art. I tried mounting rows of them on a black background. The truth is, though, I didn't find them interesting. They looked like something a computer could generate using a few simple rules. I considered hanging thousands of my geometric designs on a wall so that people could walk by and marvel at the brute effort of all that labor. But that felt less like creating art and more like giving people the opportunity to gawk at something freakish, like a large colon tumor. My doodling did feel almost cancerous at times, like when I doodled instead of working, as if this harmless pastime had metastasized into something urgent and potentially malignant.

Doodling is often defined as the kind of drawing that people do when their minds are preoccupied with thinking about something else. No one knows why people doodle. Maybe it's just a motor activity that releases tension, like fidgeting. Maybe doodling serves as a distraction for a bored person who isn't getting enough mental stimulation. Or, conversely, maybe creating a familiar doodle helps people concentrate by blocking out irrelevant and counterproductive stimuli. Different people might doodle for different reasons. A propensity for one type of doodling over another—say, interlocking triangles versus flowers or rainbows—does seem to remain fairly constant. Some self-appointed experts have claimed that doodling can reveal secret personality traits. That's why, back in the 1930s, one newspaper held a contest that promised to have a psychologist analyze a lucky winner's doodles. Thousands of people sent in their scribblings.

This contest resulted in one of the first serious studies of doodling, and it's a study that continues to get cited today, since scientists have done so little research. (A search for the word *doodling* in the biomedical research database PubMed turns up a paltry thirty-two citations, most of which have nothing to do with doodling. One, for example, is listed in the search results because a video entitled "Doodles" got used in the methods of a study called "Responses of Adult Laying Hens to Abstract Video Images Presented Repeatedly Outside the Home Cage.") The newspaper sent sacks full of self-submitted doodles to some psychiatrists who had an interest in the art of the mentally ill. In 1938, they published a report called "Spontaneous Drawings as an Approach to Some Problems of Psychopathology." They found that while only 12 percent of the doodles were "purely ornamental," by which they meant depicting a stylized pattern rather than any representational portrayal of an object or a scene, about 60 percent contained "much ornamental detail" in addition to sketches of faces or objects. Many of the doodles containing decorative detail showed "stereotypy" and repetition, both of which the researchers associated with a lowering of consciousness, similar to the cradle rocking and patting motions done to lure a baby to sleep, but also with mental disorders and mescal intoxication. They noted that doodling is a strikingly asocial activity in that no one's explicitly taught how to do it or told that they must. (Although I do remember my father telling me that he used to doodle one particular geometric design as a kid but stopped after someone told him that it looked like a swastika.) "Few people know how famous men have scribbled," the psychiatrists wrote, "and nobody tries to imitate the appearance or style of anybody else."

And yet my doodles look strikingly like George Washington's. A compendium of presidential doodles shows that when he was a boy, he inked in alternating boxes in the margins of his math assignments, creating checkerboard borders. Ulysses S. Grant drew squares and diagonal lines that look much like mine. Warren G. Harding created art deco trapezoids filled with rectangles. Herbert Hoover inked in triangles and cobweb-like swirls, and one doodle he made while being interviewed ended up on the front page of a newspaper, with a headline that read: "Everybody Does It! Scribbling President Is Held Normal." A psychologist somberly told the newspaper, "It would be significant if the President did not do this." The minds of American presidents didn't drift solely toward geometry. In addition to interlocking squares, John F. Kennedy liked to scribble sailboats. Dwight Eisenhower, a talented painter, doodled trees and bicycles and umbrellas. Ronald Reagan drew cowboys and horses. I can't understand how someone could mindlessly draw faces or animals or flowers. I sympathize more with the controlled, careful doodles of Lyndon B. Johnson, who sat in the White House filling pages of memo paper with checkerboards and repeating patterns of dashes and lines.

Such patterns go way back in history. All of the oldest-known examples of intentional human markings are geometric designs. At Blombos Cave, in South Africa, archaeologists have found pieces of red ocher engraved with lines that form crosshatches. These date back seventy thousand to one hundred thousand years ago. A report on their discovery, in the journal *Science*, described one as having "a row of cross hatching, bounded top and bottom by parallel lines and divided through the middle by a third parallel line that divides the lozenge shapes into triangles." At Diepkloof Rock Shelter, also in South Africa, ostrich eggshells engraved with similar geometric motifs date back sixty thousand years. The deliberate carvings could have held important, even sacred abstract meanings. Scientists take them as compelling evidence that early humans had evolved "modern" kinds of abstract thought.

Or, they could be mere doodles, made by some bored toolmaker who was just killing time, just as a child with a penknife might carve lines in a wooden school desk. The spirit of aimless doodling haunts the field of Paleolithic art. *The Doodle Book*, written by doodle collector and lawyer Norman Burton Uris in 1970, juxtaposed images from caves with his own doodles to illustrate that they were indistinguishable. "Today, we scribble on paper. A long time ago, people scribbled on stone," Uris wrote, carelessly dismissing long-held assumptions that cave art must have had important ritual meaning. While some ancient carvings and paintings might be great, expressive art, Uris believed that "many were doodles for doodling's sake. There are no firsthand witnesses!"

Over three decades later, one academic noted that the cave-dwellers-who-doodled hypothesis persisted in popular culture but had mostly been ignored by serious scholars. "More attention has been devoted to disparaging the theory than seriously exploring its implications in any great detail," he wrote, even though the underlying principles of doodling behavior "in modern humans, if understood, might explain several characteristics of palaeoarts, including the relative recurrence of basic forms, the nature of drawing and scratching techniques, and the common responses by rock artists to existing rock markings and geological formations." (Like cave artists reacting to existing structures, doodlers will commonly start by filling in the open spaces of printed letters or adding lines to whatever images they see before them on the page.)

Just because something *looks* like a doodle doesn't mean it couldn't also have cultural significance. The essence of doodling lies in the free-floating mental state, but someone could make exactly the same shapes with absorbed intention. In caves that are famous for their Paleolithic depictions of mammoths, lions, rhinos, and horses, archaeologists have also found zigzags, rectangles, circles, dots, and spirals. Genevieve von Petzinger, now at the University of Victoria, remembers that when she was an undergraduate taking a Paleolithic art course, she noticed that many of the photographs of painted animals also included geometric designs. These markings actually outnumbered animal depictions in Paleolithic art, but they never appeared to be the focus of attention. Her instructor said the signs hadn't been studied systematically, and von Petzinger went on to do just that.

Eventually, she determined that thirty-two geometric signs occurred repeatedly in caves across Europe, throughout the thirty-thousand-year time span of the Ice Age. "Over and over again we see different combinations of lines, dots, open angles (chevrons), ovals, hands, cruciforms (crosses), quadrangles, triangles, and circles," she wrote in her book, *The First Signs: Unlocking the Mysteries of the World's Oldest Symbols.* "This does not look like the start-up phase of a brand new invention, but more like the distribution pattern of an artistic tradition with an older, common point of origin—presumably from our ancestors' days on the African savannah." She believes that these signs might well have served as the ancient foundation of the first written languages, as each one could have encapsulated an idea in a few lines.

Humans' affinity for these particular kinds of shapes and patterns could be embedded in our biology, as the brain's neural structures seem predisposed to attend to edges, contrast, lines, and symmetry. I thought of the role of hardwired biology as I read about Jason Padgett, a self-described jock who liked to party and was indifferent to math until he got mugged outside a karaoke bar. He ended up with a head injury that, as he tells it, made him see the world in geometric patterns. In his book, *Struck by Genius*, he writes of feeling "compelled" to draw these visions, which soon were "piled high all around me in every room." Whether or not his head injury actually turned him into a math "genius" is debatable, but what seems undeniable is that Padgett's concussion resulted in his developing an irresistible urge to use a pencil and a straightedge to create fractal designs.

Some neurologists have noted that epileptics and others with atypical brain conditions can develop a burning desire to write or draw, known as hypergraphia, that makes them fill up a page, over and over. Like most symptoms of mental illness, this one seems like just an exaggerated version of normal behaviors. When I think about the time of my life when I drew geometric designs the most urgently, it was in the decade after I quit drinking alcohol but before I had children and started taking antidepressants. Back then I had a lot of free time and my alcohol-free brain hurt. I had stopped drinking because I'd tried several times to moderate my consumption, without success, and starting to live with my new husband made me realize just how much more I drank than he did. A couple of weeks after we got home from our honeymoon, I told him I was planning to have a quick drink with colleagues after work. I didn't get home until after midnight, walking in the door with a hideously scraped up face and two missing teeth, because I'd gotten into a drunken bicycle accident. The worried look on my husband's face made me realize that I had to stop drinking or I'd inevitably mess up my marriage. I didn't go to Alcoholics Anonymous or seek medical treatment. I drank lots of nonalcoholic beer, often in bed, read *Infinite Jest*, also in bed, and binged on drawing even more geometric designs than before.

After more than twenty long sober years, I read a newspaper account of a man in London who had saved hundreds of black-and-white op art drawings that had been headed for the rubbish bin after his neighbor died. This neighbor, a man named George Westren, reportedly had a troubled life and had been an alcoholic. One newspaper account said that he'd started making geometric designs after a stint in rehab. "He just wanted to draw these geometric patterns over and over again," one friend told the newspaper, and Westren had reportedly said, "Doing geometric designs got me on the straight and narrow." His designs look like my own. Maybe his brain, like mine, needed *something*, some kind of soothing, and the precision of high-contrast, repetitive, symmetrical drawing proved comforting once alcohol was no longer an option. It would be easy to say that geometric designs offer a way of creating at least one spot of order in a chaotic world. But that feels too pat, too simplistic for something with such deep roots.

Still, even though the earliest-known human etchings are geometric in nature, that doesn't necessarily mean that geometric designs are the "original," earliest form of symbolic expression or that the early humans weren't also capable of making realistic representations of what they saw all around them. Perhaps they just didn't want to. The Islamic art tradition, for example, tends toward decorative patterns rather than images of people. When reading about the supposedly "primitive" nature of geometric designs, I came across an article about Aboriginal communities of hunter-gatherers who live on the Andaman Islands, which have long been relatively isolated. Their art seems to consist solely of geometric patterns, mainly used as body makeup or as painted designs on personal items like headbands. "The most important aspect of art in the context of all Andaman hunter-forager-fishers is that they never promulgated iconographic art," one scholar wrote. "No animal, plant, or human figure features in their graphic tradition." In 1996, however, a boy named Enmay from this culture ended up in a mainland hospital for the treatment of a broken bone. Researchers found that when asked to, he easily made detailed drawings of people and animals that would look right at home scribbled in the margins of an American middle schooler's math homework.

The art of children, the art of the mentally ill, the art of supposedly "primitive" cultures—all of it has historically appealed to "real" artists and art historians who wonder about the wellspring of human artistic creativity. Children's drawings frequently look like the doodles of adults, and also like cave art. Paleolithic art scholars have measured the size of handprints in caves, along with the size of fingertips that traced lines in soft clay, and concluded that a lot of what's found inside caves might have been left there by children. Childish lines drawn in clay can be found high up in caves, as if a kid was held aloft or was sitting on a taller person's shoulders, maybe their parent's. And some handprints on cave walls were made by someone placing a tiny, toddler-sized hand against the rock and using it as a stencil, gently blowing paint through a tube all around it. This vision of prehistoric artistic life feels homey, like the cave wall was less like an altar for shamans and more like the front of the family refrigerator.

My daughter sometimes watches me on the rare occasions when I'll pull out some pens at home and work on something like a handmade birthday card. She'll get a ruler for herself so I can show her how it's done—so much for doodling not being taught. "Look," I say, "you can just make a bunch of intersecting lines and here is how you fill them in, just alternating black and white." One day, she tells me that she wants me to make the lines and she will color in the shapes. I get a giant piece of paper and sit down and try to make the triangles and squares relatively big, since she is only six years old. She hovers over the paper with a thin paintbrush and a set of watercolors. I'm interested to see what colors she picks, because except for the occasional blue or red office pen, I've never worked in anything but black and white. She boldly paints yellow right next to the blue and adds bright red and green. I wait to see if she puts in the color symmetrically or just does it randomly. She effortlessly creates a pattern of jewel-like colors with perfect radial symmetry. No one taught her this; maybe she learned it from flowers.

Meanwhile, my son scribbles geometric shapes, practicing drawing cubes by adding angled lines to squares to depict three dimensions. He tells me that his favorite shape is a "tesseract," which is the four-dimensional analogue of a cube, something my brain can barely grasp. With delight, he shows me a doodle he made on a lined index card. He drew in vertical lines to create a grid and then inked it in to look like a chessboard. Because the marker he was using started to run out of ink, the design grew fainter and fainter toward the bottom of the card, and he's pleased by this unintentional fade-out effect. I think of all the people creating checkerboard patterns, from me to George Washington to the cave people, going back tens of thousands of years, along with the equally ancient, frequently futile human drive to search the universe for meaningful patterns.

"Those designs you make, in your notebooks, to me it doesn't really seem like doodling," my husband says. I ask him, "Well, what is it then?" And he has no answer. I think of this book I am writing, the one that's now before you, another collection of black marks on white that I made in a state of half-aware compulsion, and I think also of Walt Whitman's meteors, his verses asking *What is this book? What am I myself?* I told a writer friend that my exploration of doodling felt unsatisfying, that this effort wasn't coming together in a way that made sense, and he reminded me that in an essay, unlike poetry and fiction, one can just come out and explicitly state the point of the piece, the underlying thesis or message. There's no need to be subtle or coy, he said, just let 'er rip. "But that implies that there *is* a point, that there always has to be a point that you can understand and articulate," I said. "What if there just isn't one? Like, you know, in life?"

Pondering all this awakened a long dormant urge to draw geometric designs. I once again keep a notebook open on my desk, next to my laptop, so that I can doodle as I think. After this notebook gets filled, I'll throw it on the stack with the others. I'm not sure what art is, but this robotic activity isn't it. What I'm creating just feels bloodless. I wonder if it would ever be possible to harness this kind of thoughtless activity, to transform it into something more consequential, and I think of a drawing I made a long time ago that included blood, at least symbolically. I'd been experimenting with making my geometric designs on big pieces of paper, and one day, tired of just tracing lines with a ruler, I cut out a cardboard silhouette of a sperm whale and used it to make interlocking whales embedded in my usual tangle of lines and angles and curves. In the middle of this design, I left a big empty whale-shaped void. Below and above, circles bubbled up. I inked the dark ones in red. My white whale, caught in a swirl of lines, sat in a pile of other geometric drawings for a long time. I forgot about it until I hosted a dinner party where one of the guests was a friend who'd created a popular online annotation of *Moby-Dick*. I remembered that I'd been intending to give this drawing to her, so I retrieved it from the pile and handed it over. She was surprised. And I was surprised, years later, when I went looking for a bathroom during a party in her house, and saw it framed and hanging in her bedroom, an iconic animal on the wall surrounded by geometric shapes. Once again, I'd forgotten it. Seeing it felt like hearing an echo from a past life. Whatever meaning it had, if any, was inscrutable, even to me.

Automatic beyond belief

The toaster of my childhood had, taped to its side, a yellowed newspaper clipping. It was a photograph of President Gerald Ford making toast. The picture was taken on September 5, 1974. It showed Ford inserting an English muffin into the machine, probably in slow motion for the cameras of the assembled press corps. A few days later, he pardoned Richard Nixon, and an outraged nation no longer cared that Ford was humble enough to make his own breakfast. My parents had cut out this photo and stuck it on their toaster, an almost identical model of gleaming chrome, for reasons that are both obvious and obscure.

Most of the time, our toaster lived on the kitchen countertop. At breakfast, however, it had a place of honor on the kitchen table, with its thick, fabric-covered cord snaking between two chairs to plug into a wall outlet. In one of my baby pictures, I'm sitting in a high chair pushed up to the table, next to the toaster. Looking closely, I can see the president's tiny smiling face on its side, a photograph within a photograph, a toaster on a toaster. I was born just a few months before the new president turned his toast-making into a photo op. That means I existed before Ford appeared on our toaster. I did not, however, exist before the toaster.

My mother and father received it as a wedding gift in 1959. In one vintage magazine ad for the Sunbeam Radiant Control Automatic Toaster, a tuxedoed groom and white-veiled bride are reflected in its gleaming chrome. "Automatic Beyond Belief," shouts the ad. "Bread lowers itself automatically, with no levers to push. Toast raises itself silently, without popping or banging." What's more, this modern machine could sense when the bread reached "the scientifically correct temperature for perfect toasting" and shut off the heat: "This is the entirely new toaster that has completely changed people's conception of what an automatic toaster should do!"

I had no conception of a toaster to change. When my toddler-self learned the word "toaster," that word was embodied by this particular shiny object. It looked like Sputnik, and we treated it with almost as much reverence and concern. One of my earliest memories is my father digging around in the toaster with a knife and telling me sternly, "Never do this." Toasters weren't satisfied with just burning bread, I learned—they sometimes burnt down houses. It was like living with a tiny nuclear warhead on the gold-flecked Formica.

But the way our toaster lowered the bread *was* magical, and toast was probably the first food I learned to cook. I had to drop a slice of bread from just the right height to give it a little bounce that triggered the mechanism. As the bread descended, the toaster's internal coils started to glow orange and a warm, brown smell filled the room. It was always white bread, probably Wonder Bread. I also had margarine to spread and a plastic shaker filled with cinnamon and sugar. After this morning ritual, it was time to walk up the hill to catch the school bus, or maybe it was Sunday and we'd be rushing off to church, with my father reaching over from the driver's seat to wipe cinnamon off my face, using a white cloth handkerchief (he always carried one) that he licked as I tried to squirm away from his coffee-smelling spit. I spent long hours in the pew, listening to talk of Jesus and his bread: "This is my body given for you; do this in remembrance of me."

My parents never forced me to take communion; they seemed to feel they had done their duty just by putting me in the pew. I never ate the bread of Christ. And I never took my children to church or wiped them with my own saliva. Sometimes, on Sunday mornings, I'd take them to a diner. They got chocolate chip pancakes, while I got eggs and toast. Once, as we drove to breakfast, I pointed out all the people going into churches. My son asked what people actually did at church. "They sing songs," I said, "and read their holy book. And talk about their god, and ponder the big questions like why a loving god would allow suffering in the world." My daughter, then three years old, immediately remarked from the back seat, "Maybe he allows suffering because he doesn't even exist."

When I moved away from home and had to furnish my first apartment, I went into a drugstore and bought myself a Black & Decker toaster oven. I used it to make toast, but I also would broil steaks or asparagus that I'd coated with olive oil and salt. (I never ate asparagus as a child, and the only oil I knew was Wesson oil.) This toaster oven stayed with me through my years as a single twenty-something, and when I got married, I brought it with me into my new home.

After twenty-three years, my toaster oven started to fall apart. The black plastic handle would come off in your hand, and the door wouldn't close completely. Then one day, a box arrived in the mail. It was a brand-new toaster oven, computer controlled, sent by a friend who'd recently stayed the night and had wrangled with my poor old toaster oven while heating up a bagel. Her gift was an act of love and, admittedly, a far superior toaster oven. But I knew that my parents' toaster, like their marriage, was still going strong after *fifty-eight* years. Having to replace my own felt like a failure. My toaster oven had been with me for almost my entire adult life, and I felt a little bitter that fixing it wasn't a realistic possibility. My parents' toaster, despite never breaking, was nonetheless built with the knowledge that to exist is to risk damage. The Sunbeam Corporation had even prepared a manual for any repair shops that might encounter this electric marvel, with sage advice such as "Never guess. Always investigate the source of trouble. Ask, if you do not know." What's more, "Every part, no matter how small, has a job to perform. Do not overlook the smallest detail."

I took some solace in knowing that my old Black & Decker toaster oven would soon have a new life with someone else, heroically providing toast for the heretofore toastless. I cleaned out the crumbs, shined up its metal sides, and posted a message on a local website called Trash Nothing. Previously, I'd successfully given away everything from a pile of lumber that once had been a set of bunk beds to a 50-pound bag of sand. My slightly dysfunctional toaster oven, however, attracted no takers. After a while, I put it out in front of the house on the sidewalk, with a "Free" sign. It sat there for days. I believe that eventually I took it out back to the alley and dumped it in the trash bin, but I have no memory of this. Perhaps my mind has repressed this moment because it felt so humiliating. I had not given my children a sufficiently impressive toaster. The archetypal toasting machine I had provided for their childhood was, in the end, such garbage that I could not even give it away.

It's hard to believe that anything I could impart to them will feel as solid as that mid-century toaster does to me, just as it's hard to believe that their childhoods, the only ones they will ever have, are happening right now, before my eyes, with me in the role of parent. A memory silently rises. I am around five years old, at the kitchen table of our old house, drawing cat faces on white paper plates. I announce that one is finished and bring it over to the countertop. My mother, who was likely filling the dishwasher or opening a can of corned beef hash, glances over and casually remarks, "A big part of being an artist is knowing when to stop." She could never have predicted that this particular sentence, out of all the words she has said to me, would stick in my mind. At that moment, if I had looked over at my childhood toaster, I would have seen my mother and me reflected in its shining chrome, a portrait in miniature, as beautiful and timeless as a Vermeer.

One night at dinner, I ask my children if they remember our old toaster oven, the one we relied on before the high-tech one appeared in the mail and took up residence on our counter. "Yes," my son says instantly. He is nearly eight years old. "Do you miss it?" I ask him. He thinks about it. "I don't really remember what it was like," he finally says. "I just know it wasn't as good as the one we have now."

"Wasn't our old one just a toaster?" asks my husband. "Or was it a toaster oven?"

I stare at him. How could this man live with something for two decades, interacting with it almost daily, and yet fail to observe its basic nature? I remind him that in our marriage, we have never had a toaster, only a toaster *oven*. My son asks, "What's the difference?"

This startles me. I realize that I have been raising my children in complete ignorance of the meaning of the word *toaster*. They are no doubt learning other things, things I never meant to teach them. Now, though, I carefully define the word *toaster*, explaining that it usually has one or two slits for inserting vertical slices of bread. "All it can do is make toast," I say. "A toaster oven, in contrast, is like a tiny oven. You can make other stuff in it too, like reheat pizza."

My daughter is chewing on a piece of naan, a type of bread that I first experienced in college. She swallows, takes a drink of water, and says, "I think we should get a toaster."

I wonder if she is remembering getting burned. She was reaching into our new toaster oven with a fork, going after a piece of bread toward the back. Watching, I could see that if she wasn't careful, the back of her hand might touch the hot metal edge at the top of the oven's opening. Inwardly, I agonized over whether I should say something, to warn her, but stayed quiet. Then my daughter gave a little yelp and jerked her hand away. Far in the future, when she is the same age I am now, my daughter will not recall my saying "Never do this" about forks and toasters. Instead, she will remember other words, ones I said without thinking, words that, unbeknownst to me, had just the right weight to activate some transformative process deep inside.

So when she asks for a toaster, I tell her we can think about it. In truth, I know there is only one toaster I want and only one toaster we will ever have. Years from now, when both my parents have died, and me and my brother have to clean out their farmhouse, I will bring the old Sunbeam home. The last time I saw it, it was missing its picture of President Ford, and I'm not sure why. I feel certain that my mother wouldn't have thrown something so precious away. I'll find the old photo tucked into a kitchen drawer and reattach it with new tape. I'll plug the toaster in, drop in a slice of bread, and watch it slowly descend as the heating elements switch on. The kitchen will fill with a familiar smell. I am sure this toaster will be working then. This toaster existed before me and I believe it will exist after me. I hope that my daughter truly wants a toaster. I hope that, despite all my failings as a parent, she will learn, through experience, the meaning of the word. If my children choose to carry this toaster forward, I believe it could just keep on going. As it was in the beginning, it now and ever shall be, doing its one small job faithfully and well, lifting up the golden bread as a silent offering to the heavens, forever and forever and forever, amen.

A moment of silence

Our interview is over and you are relieved. You had initially been nervous, but it turned out okay, you think. You feel you made your key points well, and I seemed to get your point of view. You liked some of my questions, and we joked around more than you were expecting. It was even kind of fun. You relax and shift in your seat, watching as I press a button on my audio recorder. You think I've turned it off, but instead I say, "There's one more thing I need to do, and it is the weirdest part of my job."

Every time I do an interview, I explain, I have to capture the ambient background noise. The audio producers need it to help smooth out transitions between the taped bits that I record in the super-silent, soundproofed studio and the tape that I record in the outside world. "I know the ambient sound here may not seem dramatic, this isn't a rainforest, with monkeys and birds screeching," I say, gesturing around your office.

"You don't have a bunch of bees buzzing here, or anything like that. But nonetheless, this place has its own distinctive sound. I need us to sit here for at least a full minute while I record it."

You nod. This doesn't seem like a high-pressure task. You don't realize that we've just started what will be, to me, the most revealing part of our meeting, although I will never include it in any story I write. "You're about to find out how long a minute really is, if you don't already know," I say in a joking tone, even though I mean this as a warning. "Ready? Starting . . . now."

The numbers on my recorder start to tick up. I watch them, but secretly I also watch you. I want to see if you're one of those people who can't stop looking at your computer, or your phone, or the papers in front of you. Now that your time to talk is over, will you shift your attention to these items, as if my presence didn't matter? Or perhaps the reverse is true—perhaps you are so solicitous of me and my needs that you become hyperconscious of your own existence, barely daring to breathe—the kind of person who will cringe apologetically when you lean back and your chair squeaks, prompting you to silently mouth the word "sorry!"

Or maybe you're my favorite kind of person, the curious kind, the kind who is intrigued by this unexpected experiment. Informed that your office contains something that's worth capturing, something you've never appreciated or thought about, your instinctive reaction is an urgent desire to experience it fully. Even without looking directly at you, I can sense your intense focus as you tune in to once-familiar surroundings. You want to hear whatever there is to hear. No matter what else has transpired between us, I feel a rush of affection for you.

The air conditioner hums. The computer's fan whirs. A car passes by on the road outside. Somewhere down the hall, a door clicks shut. Sudden laughter comes through the wall and we both smile. I might glance up at you. We have become co-conspirators, spies.

A minute, though, is a long time. After about thirty seconds, quiet attention can begin to transform into a prayer-like state of reverence. We are not two people on the subway, pursuing separate missions amid a larger crowd. We're alone together in the same confined space. Usually when you sit like this, the person in front of you would be your priest, or maybe your therapist—some kind of confessor, anyway. We're edging into a situation that's so potentially intimate, I won't even look at you indirectly. This is when I find out how close to the edge you live, how easily you can slip over into another way of being, and whether you're willing to do this in the presence of another. The not-speaking between us grows ever more pressing. You will either give in to this feeling or resist it, with your eyes helplessly wandering over your desk or up to the clock. I tend to go along with whatever emotions burble up, but then, I've been doing this for almost two decades. I knew this would be happening. It's sort of unfair to spring this on you.

This part of our meeting will leave almost no trace. No one is taking notes. The hushed *shhhhhh* that I'll hear when I listen to this recording will be your office's air handling system. As we sit, part of me idly wonders how many hours of my life have been spent like this, waiting in silent contemplation while recording fans and forced air.

One time I interviewed someone about Dorothy Parker, standing next to the writer's grave, and halfway through our conversation, I registered that a loud mechanical humming had stopped. It was a nearby building's air conditioner, and I realized that I'd need to get audio of this machine "in the clear." After the interview was over, I sat out there by myself with my recorder, waiting for the humming to switch back on. I waited for more than an hour, hanging out with the cremated remains of Dorothy Parker, imagining what she'd have to say about all this.

But here you are, very much alive. You and I were talking, like people normally do, and now together we share this quiet, which people do not normally do. It means something to me, but I have no way of knowing what it means to you, if anything, because we will never, ever speak of this. The seconds tick on and on. When the timer reaches a minute, sometimes I let the silence continue. I tell myself it's good to get abundant ambient sound, and it never hurts to have extra. The truth is that if you are listening intently with me, I won't want this to end. And every way of ending it feels awkward. I would express gratitude, but maybe this act of listening didn't feel all that special or revealing to you. Or, if it did, maybe you resent the exposure. You only agreed to an interview, not existential communion over the hissing of your radiator.

So, usually, I wimp out and make a joke. "That was a minute," I'll say, switching off my recorder with a flourish, as if to underline that our work together is now truly done. "See, sixty seconds is like an eternity! So even though this radio piece has to be fairly short, it can cover a lot of ground." We both exhale and grin. Sometimes I'll teasingly add, "If you want, I'll give you a copy of this audio so that you can listen to the oh-so-compelling sounds of your office when you're off having vacation on the beach and wishing you were here." This gives you an excuse to laugh, to release the tension, and to say something like, "Oh, absolutely, send it to me, that sounds great!" I'll never send you the audio, because I know you don't really want it. Our moment of silence is over.

Part III

My eugenics project

Early in our marriage, my husband and I idly looked at skulls with bullet holes, Civil War–era amputation saws, and obsolete X-ray equipment. We were visiting the National Museum of Health and Medicine, a collection of historical artifacts displayed at the Walter Reed Army Medical Center. All of it was interesting, in an intellectual sort of way, but then I saw something that made me freeze. Displayed at eye level, behind glass, was a bloated, dark red organ, studded with gray cysts. The cysts looked like opaque marbles. There were so many that they'd enlarged the kidney to the size of a football. Next to it, for comparison, was a healthy kidney, much smaller, smooth and pink.

"Look," I nudged my husband. Together, we stared. I tried to imagine a monstrous kidney just like it—two of them, actually—inside my husband's body, just inches away from my arm that slipped protectively around his waist.

His cyst-riddled, gradually failing kidneys were not something separate from us, an apparition we could back away from and leave behind, like that ghastly kidney on display. My husband's kidneys *were* him, or at least an essential part of him. My feelings about his kidneys set off a seemingly inexorable progression of morally questionable decisions in our marriage that still makes me queasy. We called it "The Project." In my mind, I often added another word. *Eugenics.* It was our eugenics project.

Truthfully, it was *my* eugenics project. "I think it was a mistake," my husband told me, years after it was all over. His certainty had never wavered. Despite going along with everything, participating fully in The Project, he had been against it from the start. And when I recently asked if he remembered encountering the gruesome polycystic kidney at that museum, the one I can close my eyes and see, he had no idea what I was talking about.

The two of us were like the personification of kidneys as described by the Talmud, which states that "a person has two kidneys, one of which counsels him to do good, and the other counsels him to do evil." Ancient writings like this one evidence no awareness that the kidneys did anything as mundane as filtering the blood and making urine. The kidneys, hidden as they were deep within the body, obviously spent their time pondering profound moral questions in order to advise the heart. And the Bible repeatedly mentions the heart and kidneys, together, as the essential organs that God checks out to render His own judgment: "I, God, probe the heart, and examine the kidneys, and repay each man according to his ways, with the fruit of his deeds."

That vision probably would have sounded about right to the ancient Egyptians, who hollowed out bodies for mummification but left the heart and kidneys inside. Their guides to the afterlife, such as the Book of the Dead, suggest that the kidneys were involved as the gods assessed the heart to render their final verdict on a person's life. This all-important event took place in what's often called the Hall of Truth, even though it's more properly translated as the Hall of *Two* Truths, given that a person could be judged either righteous or unworthy. The god Osiris would place a person's heart on a set of golden scales, to see if it was lighter than the sacred ostrich feather of the goddess Ma'at. Exactly what the kidneys did at this crucial moment isn't described in the mortuary texts, but they do include this plaintive cry: "Homage to thee, O my heart! Homage to you, O my kidneys!" The outcome of the weighing would be duly recorded by a waiting scribe. This was Thoth, the inventor of writing and the god of magic and science. If the scales revealed that the heart was light, indicating a well-balanced life, then Thoth wrote that the deceased could continue on to dwell in paradise. Heavy hearts, however, got thrown to the floor to be gobbled by Ammit, a demon whose body was part lion, part hippopotamus, and part crocodile. Once she devoured a heart, that was it. End of story. As one explanation of these beliefs put it, "There was no 'hell' for the ancient Egyptians; their 'fate worse than death' was non-existence."

My own heart got mixed up with kidneys when I was twenty-three years old. That's when I fell in love with the man who would become my husband. Soon after we started dating, we were lazing around in bed one morning, talking about this and that, when we got onto the subject of alternative medicine. He mentioned that he once saw an acupuncturist who stuck needles in his back and gave him "kidney pills."

"Kidney pills," I repeated. "You mean, shaped like kidneys?"

No, he told me, the pills looked like black peppercorns. They were *for* his kidneys, because he had kidney trouble.

"What kind of kidney trouble?" I asked, recalling him mentioning that his mother had needed a kidney transplant.

Polycystic kidney disease, he said. It makes kidneys grow more and more fluid-filled cysts, until eventually the organs stop working. He'd been diagnosed when he was about twenty-four years old; he'd had some blood in his urine and a scan of his kidneys revealed a lot of cysts. He seemed casual about the whole thing. He noted that medical science improves constantly. This disease was genetic, obviously, but he didn't think the risk of passing it on was any reason not to have children.

Later, in my journal, I wrote that I had been "a little shocked" but that, upon reflection, it didn't matter: "If it works out so that we want to stay together forever, that would be a wonderful thing, even if one of us died or had a disastrous health problem. It's not as if I give my lover a eugenics checklist that he has to meet."

What I did not write, even in my private notes, was that the intensity of my reaction had startled me. This wasn't some devastating genetic disease like Tay Sachs, one that guaranteed a painful death in childhood. But any kid unlucky enough to inherit this disease gene would almost certainly need a kidney transplant in adulthood—sometimes *early* in adulthood.

I didn't think that an organ transplant at the age of thirty or forty, and then years of taking drugs to suppress the immune system, was anything to just shrug off. The thought of that happening to my theoretical children disturbed me, despite the fact that it was a treatable, adult-onset illness, with my paramour right there in bed beside me to prove that polycystic kidney disease did not prevent a happy life. I felt unsettled. And ashamed of feeling unsettled. Did some part of me recoil from possibly sullying my genetic line with a substandard kidney gene? Was I secretly, at my core, a eugenicist?

....

The word *eugenics* has a sinister sound that conjures up thoughts of Nazis. The idea of trying to deliberately improve humanity with scientific breeding was around for decades before the Nazis came to power, however, and I knew that the Nazis had drawn inspiration from prominent thinkers in the United States. The English scientist Francis Galton—a cousin of Charles Darwin—may have first coined the word *eugenics* in 1883, from Greek roots meaning "well born," but his ideas took off across the Atlantic.

At the center of it all was the Eugenics Record Office at Cold Spring Harbor in Long Island, New York. It was part of a genetics research group founded by a Harvard-trained biologist named Charles Davenport, who had convinced the Carnegie Institution of Washington to set up a center for the experimental study of evolution. Much of its work focused on plants and animals. But the goal of its Eugenics Record Office, created in 1910, was to improve "the natural, physical, mental, and temperamental qualities of the human family." This progressive-seeming vision attracted additional wealthy funders such as the Rockefellers, and the eugenics office trained workers to travel around the country and collect detailed histories of purportedly superior or inferior families. The workers filled filing cabinets with elaborate pedigree charts. "Science cannot experiment with human beings," explained Davenport. "It desires merely to learn and publish the actual results of man's experiments on himself. Every marriage is an experiment in heredity."

All kinds of traits had a genetic basis, according to the eugenicists—everything from medical conditions like albinism and hemophilia to far more subjective qualities like "feeblemindedness" or "shiftlessness." It's easy to laugh at a family history purporting to document an inherited predilection for "boat building." But the values and assumptions of eugenics seemed like common sense to many Americans, including luminaries like Alexander Graham Bell, Margaret Sanger, and Theodore Roosevelt. The former president wrote to Davenport in 1913, saying that "society has no business to permit degenerates to reproduce their kind. It is really extraordinary that our people refuse to apply to human beings such elementary knowledge as every successful farmer is obliged to apply to his own stock breeding." This comparison to farming and livestock was made explicit at state agricultural fairs, where the American Eugenics Society set up exhibits. One display showed how fur color got inherited in guinea pigs and warned that "unfit human traits such as feeblemindedness, epilepsy, criminality, insanity, alcoholism, pauperism and many others run in families and are inherited in exactly the same way as color in guinea pigs." This suspect educational material was accompanied by competitions dubbed "Fitter Families for Future Firesides," created with Davenport's assistance. Parents put their children forward for hours of inspection and judging by eugenicists, just as they might go for a blue ribbon at the fair by submitting an especially fine turnip.

The winners of these "Fitter Families" contests were invariably white, educated, well-off Protestants. Eugenics spread a thin veneer of science over the sweeping class and race prejudices of the day. The "better" classes of society would be overrun, eugenicists warned, by dark-skinned hordes indiscriminately birthing degenerates—and just imagine the tax burden of having to support generations of genetically predestined paupers! Congress adopted tough immigration restrictions after the superintendent of the Eugenics Record Office testified that immigrants from places other than Northern Europe were more likely to be "socially inadequate." And in 1907, Indiana passed the world's first law allowing the government to sterilize "confirmed criminals, idiots, imbeciles and rapists" so that they couldn't pass on defective genes. Thirty states followed. Throughout the twentieth century, doctors forcibly sterilized around sixty thousand Americans.

One of them was Carrie Buck, whose tragic story encapsulates the whole wretched business. Doctors at an institution called the Virginia State Colony for Epileptics and Feebleminded had been searching for a test case to prove the legality of compulsory eugenic sterilization. Seventeen-year-old Carrie Buck, who had just arrived at the asylum, seemed to fit the bill: her mother Emma was already living there, and Carrie Buck also had an out-of-wedlock infant named Vivian. This seven-month-old baby was examined by a social worker who concluded, "There is a look about it that is not quite normal, but just what it is, I can't tell." That pathetically limited assessment, plus testimony from a Eugenics Record Office expert on the wisdom of cutting Carrie Buck's fallopian tubes, was enough to convince almost everyone, including the US Supreme Court. In the famous 1927 decision *Buck v. Bell*, Justice Oliver Wendell Holmes Jr. declared: "Three generations of imbeciles are enough."

Historian Paul A. Lombardo, in his book *Three Generations, No Imbeciles*, has pointed out that nothing was ever wrong with the Bucks except that they were poor and female. Carrie Buck had been in foster care because her mother couldn't afford to support her. Her foster parents only foisted her off onto the asylum because she was pregnant, and the baby's father was her foster mother's nephew. Later in life, Carrie Buck told reporters that he "forced himself on me . . . he took advantage of me." Lombardo's research shows that the attorney who had supposedly defended Carrie Buck was a supporter of eugenics and secretly in cahoots with his buddies at the asylum who wanted to sterilize her. This lawyer called no witnesses and he failed, Lombardo wrote, "because he intended to fail." Doctors sterilized not only Carrie Buck but also her younger sister, believing that she, too, was feebleminded and "sexually delinquent." Her name was Doris, and she thought the surgery was for her appendix. After she got married, Doris spent years trying to get pregnant before finding out the truth. And Carrie Buck's baby, Vivian, who reportedly looked "not quite normal"? She died at the age of eight from an infection, but before that she went to school. She got fine grades on her report cards.

Plenty of people back then saw eugenics clearly. Attorney Clarence Darrow flatly called eugenics a cult in 1926 and mocked its proponents for thinking they had any idea how to improve the human race. "I, for one, am alarmed at the conceit and sureness of the advocates of this new dream," Darrow wrote. "I shudder at their ruthlessness in meddling with life. I resent their egoistic and stern righteousness. I shrink from their judgment of their fellows." What about kindness and understanding, Darrow wondered, adding that "talk about breeding for intellect, in the present state of scientific knowledge and data, is nothing short of absurd." The state of scientific knowledge was extremely limited. Eugenicists frequently cited Gregor Mendel's groundbreaking nineteenth-century studies in pea plants that established some of the basic rules of heredity, but no one had any understanding of what physically got passed from parent to child to produce particular traits; eugenicists could only refer vaguely to "tainted germ plasm."

Scientists who were trying to figure out what heredity truly was, like Thomas Hunt Morgan, pulled away from any association with eugenics. Morgan studied mutant fruit flies in his renowned "Fly Room" at Columbia University and won the Nobel Prize for showing that units of heredity, or genes, must be carried by the cells' chromosomes. He wrote in 1925 that "in the case of man's physical defects, there are a few extremely abnormal conditions where the evidence indicates that something is inherited, but even here there is much that is obscure." With something as difficult to define as feeblemindedness, Morgan wrote, it was hard to know what was even being measured, to say nothing of all the possible nongenetic influences: "It is obvious that these groups of individuals have lived under demoralizing social conditions that might swamp a family of average persons." Biologist Raymond Pearl was even harsher, calling eugenics "a mingled mess of ill-grounded and uncritical sociology, economics, anthropology, and politics, full of emotional appeals to class and race prejudices, solemnly put forth as science, and unfortunately accepted as such by the general public."

In 1935, the Carnegie Institution of Washington asked a group of independent scientists to visit the Eugenics Record Office in Cold Spring Harbor and review its activities. They concluded that almost all of its carefully cataloged pedigrees were worthless, or, as they put it, "unsatisfactory for the study of human genetics." By the end of 1939, the Eugenics Records Office got shut down. Soon after, in Germany, Adolf Hitler showed the world exactly what it looked like to implement a program of eugenics on a large scale. The idea fell out of favor.

"By the time I first came to Cold Spring Harbor for the summer of 1948, accompanying my PhD supervisor," biologist James Watson later wrote, "the Eugenics Record Office had been virtually expunged from its consciousness." The labs in Cold Spring Harbor were instead focused on trying to understand the fundamental nature of genes. In 1953, Watson and Francis Crick published their paper describing the structure of DNA, noting that its unusual features "are of considerable biological interest." DNA, they pointed out, was made of four chemical building blocks that could get strung together in endless variations, like a code. And DNA also had a spiral-staircase-like structure that meant it could unwind into two halves, each of which could act like a template for making a copy of the original. By 1966, other researchers had cracked this genetic code, showing how combinations of DNA's building blocks could hold the instructions for making proteins.

Researchers began developing the first genetic tests in the 1980s, for diseases like Huntington's and cystic fibrosis, and Watson pushed for the federal government to support a controversial effort to map all human genes. At the time, he was serving as director of the Cold Spring Harbor Laboratory, and when he also became the first leader of the Human Genome Project, he made the unprecedented announcement that 3 percent of its funding would go to studying its ethical, social, and legal implications. "In putting ethics so soon into the genome agenda, I was responding to my own personal fear that all too soon critics of the Genome Project would point out that I was a representative of the Cold Spring Harbor Laboratory that once housed the controversial Eugenics Record Office," Watson wrote in an essay entitled "Genes and Politics." If he didn't focus on ethics quickly, Watson feared, that "might be falsely used as evidence that I was a closet eugenicist, having as my real long-term purpose the unambiguous identification of genes that lead to social and occupational stratification as well as to genes justifying racial discrimination." Ironically, given these concerns, racial discrimination is what ultimately forced the Cold Spring Harbor Laboratory to cut all ties with Watson, decades later, after the elderly scientist repeatedly made embarrassing public statements asserting that the intelligence of people of recent African descent was genetically inferior.

That scandal came long after the triumphant completion of the Human Genome Project, which was celebrated at the White House in June of 2000. Watson attended the event, and President Bill Clinton lauded him as a visionary. I was there, too. I was twenty-six years old, working as a reporter assigned to cover this historic occasion. "More than 1,000 researchers across six nations have revealed nearly all 3 billion letters of our miraculous genetic code," President Clinton told the assembled crowd. I earnestly scribbled his words in my notebook, thinking only of the deadline for the news article I had to write. On my left hand, I wore an opal engagement ring. My wedding was just a few months away. I'd agreed to marry the man with polycystic kidney disease, despite my eugenic fears. "Genome science will have a real impact on all our lives," the president continued, "and even more, on the lives of our children."

I knew that if I married him—this charming, intelligent, gorgeous, funny, loving, kind man—it was quite possible that we'd pass on polycystic kidney disease to our children. Inside each of his body's cells was one dominant version of a gene that caused this disease and one recessive version that did not. Whenever his body produced sperm, which contains half the genetic material needed to make a human being, each of the sperm would receive only one version of this gene: either the faulty one or the normal one. Depending entirely on which sperm ended up fertilizing the egg, a child would either get the disease gene or get lucky. That meant each time a child was conceived, the chance of passing on the disease was fifty-fifty, like a coin toss.

"I didn't know the risk was that high," my then-boyfriend,

now-husband said when I told him the odds. "I thought it was more like 25 percent."

It was early in our courtship, and we were sitting together in a bar called the 13th Floor, at the top of the old Belvedere Hotel in Baltimore. We'd just seen the movie *Gods And Monsters* and had walked together through a light snowfall, taking the elevator up to sit in this cozy spot with a view of the snow and the city lights down below. I was telling him what I'd learned about polycystic kidney disease from the Internet. I was astounded at his lack of awareness, his indifference to something that loomed so large in his family's life.

Although, in fairness, why would he bother thinking about something he couldn't change? Doctors had no treatment to prevent the growth of the innumerable cysts that were slowly obliterating his kidneys. And at that time, in 1998, there also wasn't any simple DNA test for this disease. The disorder could be caused by any one of many different mutations, most of which were a mystery, in more than one gene. Back then, even without knowing the exact genetic typo, doctors could sometimes trace the overall pattern of heredity within a family by using well-established DNA markers and blood samples from several family members. This kind of genetic testing was expensive, though, and not a sure thing. It was generally only done to establish that a family member who wanted to donate a kidney didn't have the disease or "when the outcome of a pregnancy would be altered if a positive diagnosis were made in the fetus," as the Polycystic Kidney Disease Foundation's website delicately put it.

We talked about all this over pizza and two glasses of wine.

He told me that he was pro-choice. But even so, he couldn't see testing a fetus for the same disease he had and, if it tested positive, ending it with an abortion.

And I agreed. Because there he was, thirty-two years old, and he seemed fine! He had recently hiked the 2,190 miles of the Appalachian Trail. He played tennis; he jogged. Other than the minor bout of bloody urine that led him to have his kidneys scanned, he had no sign of illness. And that's typical for this disease. The cyst formation usually takes decades to become disabling. Many people don't even get diagnosed until their thirties or forties—often long after they've had kids and potentially passed the disease gene on.

That was what happened to his mother. She grew up in Brooklyn, New York, where her dad owned an auto parts store. When she was thirteen years old, in 1954, her father died of kidney failure. Doctors had no way to save him; it would be more than a decade before the opening of the first dialysis clinic with machines that could clean people's blood. And although scientists had been trying human kidney transplants from cadavers as far back as the 1930s, the recipient's immune system always attacked the foreign organ. (A notable exception came the same year that her father died, when doctors transplanted a kidney from one identical twin into the other.)

When she got married in June of 1961, she had no idea she had kidney disease. Not too long after that, in 1963, scientists met in Washington, DC, for a conference on kidney transplantation. The overall picture was demoralizing; less than 10 percent of several hundred kidney transplant recipients had survived for more than three months. But at that meeting, a researcher named Thomas Starzl announced that he'd been having much better success with a combination of drugs, including corticosteroids. "The outlook for kidney transplantation was completely changed by Starzl's report," according to one account. Researchers kept at it, trying different immunosuppressants, and in 1983 the Food and Drug Administration approved the powerful antirejection drug cyclosporine, ushering in a new era for organ transplants.

That was the year my husband's mother got one. She'd had a urinary tract infection that turned out to be related to infected cysts, and she learned that she had the disease that had killed her dad. Her own kids were teenagers, and she was stuck in the hospital, coping with infections and rejection episodes. Medical science ultimately gave her two more decades of life than her father had. But she was often in pain, or sick. The immunosuppressant drugs had serious side effects, like weakening her bones. The first time I went to my husband's childhood home, his mother's leg was in a huge cast and she could barely get out of bed. She'd fallen down while on a bike ride and had broken her femur.

One day, a few years after our wedding, my office phone rang. It was my mother-in-law. She'd never called me at work before. She wanted to know if she could ask a personal question about her son. "Um, you can ask, but I might not answer," I said, trying to prepare myself for whatever might be coming. It turned out that she just wanted to know what was going on with his kidneys. She thought he had told her that he needed a kidney transplant, that he was in renal failure. "Oh, no, not yet," I said. I told her his most recent blood test results and assured her that he was fine. It sounded, over the phone, as if she began weeping with relief. She apologized for bothering me at work, for being so silly. "I get confused," she said, explaining that it was the painkillers she was taking. Maybe she'd had a dream that she thought was reality.

What touched me was that she hadn't called my husband. I wondered if she found this subject difficult to broach with him; I wondered if she was too fearful, or maybe ashamed. I remember visiting her toward the end of her life. At that point she was in a wheelchair. She'd just miraculously survived meningitis and almost immediately after that been diagnosed with a cancer that seemed to be related to long-term immunosuppression. We were alone in the house and I was reading the newspaper in the kitchen when, with some difficulty, she wheeled herself in. "I'm scared to die," she said, getting right to the point. I put down the paper and said, "I don't blame you."

When she did die, I wrote her obituary. She'd asked me to do it on her deathbed, speaking through an oxygen mask. "I'll write it," I said, "but I hope I don't have to for a long time." I don't know why I said that. I knew, and she knew, that this time she wasn't going to pull through. "Put in there that I *tried*," she said, working hard to breathe. "I want that to be my epitaph: She *tried*." I didn't understand, but I included that in her obituary. It felt like the most difficult writing assignment ever, and one that I failed.

That same month, October of 2004, something else appeared in print. A medical journal published the first report of embryos created through in vitro fertilization (IVF) so that they could be tested for polycystic kidney disease. A woman with the disease, whose father and brother also had it, had gone to a fertility clinic in Chicago because she wanted to see if it was possible to create a disease-free pregnancy. Even though the doctors didn't know her family's exact disease-causing mutation, they were able to take DNA samples from different family members and use genetic markers to come up with a reliable test. They took eggs and sperm, put them together to make fourteen embryos, and plucked a single cell from each one to test its DNA. Three of the embryos showed no sign of the disease gene. A doctor put them into her uterus, and she gave birth to twins.

I didn't immediately see that article. I wasn't routinely scanning the scientific literature for news of the latest advance in polycystic kidney disease. When I got married, in front of all my friends and family, I had pledged to have and hold my husband "from this day forward, for better, for worse, for richer, for poorer, in sickness and in health." To me, saying "I do" meant accepting that our children might inherit this disease. And I did.

A few years after my mother-in-law's death, however, I went to visit my gynecologist to tell her I wanted to try to have a baby. As she went over the various tests and vaccinations and vitamin recommendations, she asked about our medical histories. I told her that my husband's family had polycystic kidney disease but that it seemed like something that was difficult to test for. My gynecologist listened closely, made a notation in my medical file, then wrote a name and phone number on a prescription slip. "I think you should see a genetic counselor so you understand what's possible," she said, ripping off the slip and handing it to me. "All of this genetic stuff is developing unbelievably quickly."

The concept of genetic counseling was developed in 1947 by a researcher named Sheldon Clark Reed. He was the head of a place that was originally supposed to be a eugenics center, the Dight Institute for Human Genetics at the University of Minnesota. It was created with a bequest from Charles Dight, an ardent eugenicist who had corresponded with Davenport and Hitler, campaigned for the expansion of forced sterilization, and published pamphlets like *Human Thoroughbreds—Why Not?* One of the first things Reed did after arriving at the institute was to have a bunch of files transferred there from the defunct Eugenics Record Office at Cold Spring Harbor Laboratory. Reed searched through its tens of thousands of pedigrees to look for anything useful, like families with real genetic disorders, and he often wrote about eugenics in a positive way. Even so, in those post-Nazi days, he understood the need to tread carefully. "It is my impression that my practice of divorcing the two concepts of eugenics and genetic counseling contributed to the rapid growth of genetic counseling," wrote Reed. "Genetic counseling would have been rejected, in all probability, if it had been presented as a technique of eugenics." He once remarked in a lecture that there was "no great philosophical distinction" between eugenics and human genetics but that avoiding the former term might be "financially and politically expedient." Despite his acceptance of eugenic goals, however, Reed maintained that doctors and other experts should give people objective medical facts and let them come to their own decisions about making babies. Genetic information should be delivered, Reed said, in a way that was "compassionate, clear, relaxed, and without a sales pitch."

That's pretty much what we got when my husband and I went to see the genetic counselor. The appointment lasted for hours. We were there to talk about kidney disease, but the counselor started out by sitting us down and asking detailed questions about our families. "What I will do is document your family history, draw a picture of it," she said. "This is a pedigree, or a family history chart. Circles are females, squares are males." As we dutifully recited every case of heart disease, cancer, renal failure, or general weirdness going back several generations, the counselor marked it all down, using a standardized set of symbols and notations that she may or may not have known was codified by Davenport in his 1911 treatise *Heredity in Relation to Eugenics*. She seemed especially intrigued by the pattern of breast and ovarian cancer on my mother's side of the family, to the point where I grew impatient—we were there to talk about my husband's genetic disease, not some theoretical familial cancer risk in my far-flung relations.

Finally, we got into the nitty-gritty of polycystic kidney disease genetics. We learned that a commercial lab could now read the individual genetic letters in my husband's DNA, and the chances were extremely good that they'd find the disease-causing mutation. We could get a made-to-order test and use it to screen either embryos or fetuses—although the test for fetuses couldn't happen until the pregnancy reached around eleven weeks. The counselor, no doubt wanting to be thorough, noted that another way to avoid this disease would be to use a sperm donor. She mentioned it briefly, in a tone of voice that suggested she knew we wouldn't be interested, and we weren't.

As for the rest of the options—well, my husband hated

them. He was against all of it. He was especially repelled by the idea of going to a fertility clinic and, in his opinion, unnecessarily handing over the most personal part of our marriage to a bunch of strangers in lab coats. His feelings about this were visceral; he said he'd rather not have children than turn to the reproductive industrial complex. He wanted to conceive children at home, in private, or not at all.

I couldn't understand why he wanted to get pregnant like we were living in the 1960s. He'd never expressed any objection to common pregnancy tests like ultrasounds, amniocentesis, or blood tests used to detect disorders like Down syndrome—another condition that is frequently compatible with living a happy life. And yet he'd inexplicably take a pass on technology that would let us have our much-desired baby while avoiding the risk of passing his family's awful kidney disease. Admittedly, it was a treatable illness, but one that had killed his grandfather and tormented his mother. And it was destroying two major organs inside his body. Didn't he have any desire to strike back? To eliminate it from all future generations?

We talked about this while making dinner, while driving to do errands, while getting dressed in the morning—and we couldn't come to any agreement. "You know, people die. All sorts of things could happen to a kid. A kid could get hit by a bus," my husband said. "Going to extreme, extreme measures to avoid one potential way that your kid could be hurt or could die is, in a way, kind of dangerous. It seems to indicate that you're not accepting that you're not in control of what's happening to your kid." Trying to pick and choose our child's genes would be a telltale sign, he said, that we had the wrong mindset about parenting, and about life.

His fatalism baffled me. "It's not like every kid gets born with a 50 percent chance of being hit by a bus! This is an entirely different kind of thing," I insisted. "This is something we know about and can prevent." It felt significant to me that all of the reproductive technologies that the doctors had offered—whether it was aborting fetuses with the disease gene or using ovary-stimulating drugs and egg harvesting to create testable embryos in a lab—would involve *my* body. And to me, it felt like a parental duty to take on a whole lot of unpleasant medical interventions to ensure that our child would never have to.

I felt this way even though the logical part of my brain knew that this was like believing in magic. I could never take this physical burden from an imaginary child and put it on myself. If I offered up my body to the altar of medicine, the whole point would be to produce a child that didn't inherit this stupid gene at conception. Those children are exactly the ones who would have never faced any kidney-related difficulty; I would be giving that child nothing that it didn't already have from the get-go. The only potential children whose future could have included kidney-related travails, the kind of troubles that I longed to spare them, would be the very ones that my eugenic efforts wouldn't allow to exist. It's hard to see how prohibiting someone's existence can be a form of love, unless it's done to prevent constant, unremitting pain—which was not the case here. Most mortal beings assume that existing is almost always better than not. But then we are biased. Ask my husband if he prefers existing with polycystic kidney disease to not existing, and naturally he is grateful that he gets to experience the universe, including the chance to ponder that question. The opinions of those who don't exist cannot be taken into account. They do not exist. Their condition, non-existence, is the fate that the ancient Egyptians saw as being worse than death.

"Maybe all this would be easier if you had already had your kidney transplant," I told my husband one day, starting to cry. "Maybe that would make things easier for me. Because the way it is now, I get scared about what's going to happen to you. You assume everything is going to be fine, and maybe you have to so that you can get through this, but maybe it's harder for the person who just has to watch." How long did he think his kidneys could last, I asked—another five years? (A wildly optimistic projection, as it turned out.) I told him to picture me five years from now, caring for my post-transplant husband as he recuperates from surgery and pops handfuls of immunosuppressant pills and painkillers. I'd be bringing him soup in between taking care of our brand-new infant, our darling little Francine or whatever her name would be, all the while wondering if I was watching a preview of what she would have to endure because we didn't bother to take advantage of medical science when she was conceived.

My husband said if I insisted on doing something to prevent passing on this disease, he might be okay with getting pregnant the old-fashioned way and then testing the fetus, because that would not involve IVF, which is what he most objected to.

"But if we did go the screen-and-have-an-abortion route," he said, "projecting myself into the future, it seems almost for sure that when we had this eleven- or twelve-week-old fetus, that even if we learned it had polycystic kidney disease, we'd say, 'Well, we still love our little fetus and it's really not that big a deal to have polycystic kidney disease, so we're not going to abort our little fetus.' "

"I could abort a fetus with polycystic kidney disease," I warned him.

"You say that now," he said. "I mean, sure, you could *force* yourself to do anything. But I don't think we'd want to. You know, twelve weeks, when they could get test results back, that's almost three months. We'd be pretty far along in the process. I think we probably wouldn't do it."

If we took that path, I cautioned, it would have to be with the expectation that we'd do what we agreed. I wouldn't let myself get emotionally attached. "It wouldn't be like a normal pregnancy. It would be like a probation pregnancy," I said. "I know that sounds horrible."

It did sound horrible. It felt shameful. I didn't want to tell anyone that we were contemplating this. Still, I made another appointment with my gynecologist and explained that my husband and I thought we might try getting pregnant and testing the fetus. I wanted to know if she would participate—to put it bluntly, if she would end a pregnancy because it had tested positive for a treatable disease that didn't cause symptoms until adulthood. She replied that IVF with embryo testing made more sense to her, but we'd clearly thought about it a lot. She was my doctor, she said. She would give me the medical care that I wanted.

This wasn't what I wanted. Before I got married, I'd sat in a bar, gazed into my lover's eyes, and agreed unequivocally with his dismissal of abortion as a way to stop this genetic disease. What had changed? Were my feelings altered by the knowledge that it was now possible to create embryos in the lab and then selectively start a pregnancy with unaffected ones? Had my views shifted after watching the results of my husband's kidney function tests get worse and worse, year after year? Was it the experience of seeing his mother in the hospital as she lay dying, struggling to breathe, and then having to write her obituary? Or maybe it was that day his mother called me on the phone at work, weeping over the thought of her son in kidney failure, and realizing that, before too long, the mother facing that pain could be me?

That night, I told my husband that my gynecologist had no objections. "It's not the choice that I would make," he said once again, but he wasn't going to stand in my way. Our discussions had led us to a perverse compromise: I'd get to act to prevent the transmission of this inherited kidney disease, and he'd get to avoid IVF. Yet this meant we both had to agree to something that we didn't want to do.

"For me it's basically just a matter of time and a little bit of hassle," said my husband.

"It could be beyond a little bit of hassle," I reminded him. "It could be that I get pregnant—and the thing has the disease, so we have an abortion. I get pregnant again—and the thing has the disease, so we have an abortion." I inwardly noted that I had referred to our deeply desired future baby as a "thing."

"That's the time aspect," my husband said, seemingly

unperturbed. He wanted to have a baby and was concerned that this effort would make it take longer to start a family.

"But it's going to be emotionally difficult too," I pointed out. "Aren't you going to be sad if I have an abortion?"

"It's going to be disappointing," he allowed.

"You're not going to be mad at me?" I asked.

"No," he said.

"For killing your baby?" I said, wanting to push this as far as it might need to go.

"No," he said again, with a kind of rueful half laugh.

"Are you saying all this," I asked, "because you think if I got pregnant then when the time came to do a termination that I wouldn't be able to do it? And this is a way of getting around the whole thing?"

"No," said my husband. "No, I'm not playing little games like that."

And so The Project—my eugenics project—began.

Some would argue that it's unfair, even sensationalistic, to use the word *eugenics*, which is so loaded with negative connotations—Nazis! Forced sterilization!—that it can shut down rational thought. That may be one reason why these three potent syllables often get thrown around by those who object to any genetic testing of fetuses or embryos. Even though I didn't feel that my husband and I had stepped out onto a slippery slope that would inevitably end with a blonde, blue-eyed master race of superhumans, we were taking fairly extreme actions to try to better the genes of future generations. What is that if not eugenics?

"Past eugenics is assumed to be something despicable that ought not to be repeated, but those who participate in debates on the ethics of reproductive technologies and practices often fail to explicitly refer to what was wrong with eugenics and why," notes Giulia Cavaliere in her article "Looking into the Shadow of Eugenics." To her mind, the eugenics apocalypse had four horsemen: ludicrously inaccurate scientific claims, reliance on coercion and force, blatant race and class prejudice, and an unhealthy obsession with the pursuit of perfection.

Well, then. My effort to avoid polycystic kidney disease was based on solid science. And no one was *forcing* me to do this.

But coercion can be subtle. Wasn't my judgment swayed by the mere availability of all these advanced genetic tools, by the knowledge that the medical establishment had embraced them, and by the reports of other families doing it? Wasn't my husband being coerced by me?

And let's not pretend that this whole enterprise operated in some pristine realm of pure medicine, untouched by prejudices and class-based injustice. A poor person in this country with polycystic kidney disease wouldn't have the same options that I had, because doctors generally only offer up their bevy of high-tech reproductive options to the rich. Personalized genetic tests and IVF cost serious money and usually aren't covered by insurance. What's more, someone might legitimately argue that testing of embryos and fetuses can be a form of discrimination against people with genetic disorders; such testing implies that perhaps it would be better if they had never been born. I didn't feel that way about my husband. But still. The stigmatization of the disabled or sick is ugly and real.

On the other hand, almost *every* parent tries to control their children's genetic makeup. That's true of every parent who decides to reproduce with a particular person. By selecting my husband, and not any other man out of billions of men, I had already taken a major step to determine my eventual children's genetic heritage. Maybe I didn't like one of my husband's genes, but I sure wanted the rest of them. And almost no one would object to my right to choose them.

"When reproduction occurs *au naturel*," wrote bioethicist John Robertson in his book *Children of Choice*, a person's "procreative liberty" is widely understood and accepted. He defined procreative liberty as the "freedom to decide whether or not to have offspring and to control the use of one's reproductive capacity." He felt this freedom was paramount whenever a person decides to reproduce—or to *not* reproduce—regardless of what kind of technology might be involved. "Although I conclude that few ethical or legal limits on the use of reproductive technology can be justified," Robertson wrote, "I never deny the profound ambivalence that may attend recognition of procreative autonomy."

That ambivalence comes because the genetic choices of would-be parents can seem incomprehensible to others, as all of this depends so much on a person's unique experiences. Some scholars have called this "flexible eugenics." Deaf parents may seek out a deaf sperm donor, hoping for deaf children. Parents with dwarfism may welcome children who are Little People while wanting to end pregnancies that have a more severe, deadly form of the condition. A woman with breast cancer who lost her mother to the disease might turn to IVF and discard embryos that have the familial breast cancer mutation, even if it only conveys a *predisposition* to cancer that might never actually happen. Another woman, longing for a daughter, might test embryos to make sure the doctor implants only female ones. And a couple that knows they have a risk of producing a child with a disabling genetic disease may forge ahead without any kind of testing, leaving the outcome up to their god. The father of genetic counseling, Sheldon Clark Reed, assumed that informed parents would naturally strive to have children who were "genetically normal." But what is normal? To me, normal was living life without counting down the days to total kidney failure; to my husband, having an organ transplant looming in his future was as unremarkable as drinking his morning coffee.

I was about as far removed from a eugenics victim like Carrie Buck as possible. I was rich, not poor; a volunteer, not under court order; working to have a child, not having my fertility stolen away. And the world has changed a lot since 1910, the year that saw the creation of the Eugenic Record Office. Still, if I could have traveled back in time and gone there to have a chat with Davenport—who urged couples to consult with his office—I felt certain that he would have approved of what I was up to in trying to wipe out this kidney disease gene from my future family. In his view, every person "should look forward to securing a scientific analysis of the hereditary potentialities of himself and all his descendants." He believed that "when this desire becomes general, the science of eugenics will become firmly established."

My husband went back to the genetic counselor, got some blood drawn, and filled out paperwork. His blood and $4,000 went to a genetics lab, which isolated the two genes that are often mutated in polycystic kidney disease. Lab workers looked at these stretches of DNA and checked the sequence of its four chemical building blocks, known by their initials: A, C, T, and G. By proofreading more than seventeen thousand of these chemical letters, the lab was able to discover a tiny error. Where there should have been a G, there was a T. This particular mutation had never been described in a medical journal before, so there was no ironclad proof that it was the culprit. But if a cell tried to make a protein using the genetic instructions with this typo, it seemed that the resulting protein wouldn't work right. I marveled that science had found the small genetic mistake that had such a huge impact on my husband's family.

Armed with this knowledge and assurances that the lab could test fetal DNA, we started trying to get pregnant. I took my temperature every morning and made charts to track ovulation. It wasn't romantic. It felt like a science experiment. I feared that by doing all this, we were guaranteeing that a horrible fate would await any resulting child, some ironic twist worthy of an O. Henry tale. As I told my husband, it would be like, "Oh, they tried so hard to make the perfect, perfect baby . . . and then it was hit by a meteorite."

As for the prospect of possibly having an abortion—I can't add much to the innumerable words that have already been written on this subject, except to note how often arguments over abortion veer into scenarios that involve kidney failure. "I don't want to hear how 'pro life' you are unless you've donated a kidney. Pregnancy imposes a greater risk than donating a kidney, so go out and save a life," one obstetrician wrote on Twitter, noting "it is so easy to be 'pro life' when it is theoretical and someone else's body." In the influential 1971 essay "A Defense of Abortion," philosopher Judith Jarvis Thomson used kidney failure to point out that even if the fetus has an unambiguous right to life, it doesn't necessarily follow that a woman is obligated to help it live:

Let me ask you to imagine this. You wake up in the morning and find yourself back to back in bed with an unconscious violinist. A famous unconscious violinist. He has been found to have a fatal kidney ailment, and the Society of Music Lovers has canvassed all the available medical records and found that you alone have the right blood type to help. They have therefore kidnapped you, and last night the violinist's circulatory system was plugged into yours, so that your kidneys can be used to extract poisons from his blood as well as your own. The director of the hospital now tells you, "Look, we're sorry the Society of Music Lovers did this to you—we would never have permitted it if we had known. But still, they did it, and the violinist is now plugged into you. To unplug you would be to kill him. But never mind, it's only for nine months. By then he will have recovered from his ailment, and can safely be unplugged from you." Is it morally incumbent on you to accede to this situation?

Her answer was no. It might be a lovely, generous thing to do for a fellow human being, but no law should require it:

the law does not force us to be Good Samaritans. This was the first of a series of mind-bending hypotheticals in her article, which I read in college. Her conclusions about abortion resonated with my own instincts, because they weren't simplistic. While strongly endorsing a woman's right to control her own body, they allowed for the possibility that in some cases, the decision to abort a fetus might be morally indecent. Her example of that was if a woman with a well-advanced pregnancy decided to abort to avoid the inconvenience of having to postpone a European vacation. Her article didn't say anything, though, about the moral decency of deliberately conjuring up a fetus with the intention of possibly killing it for having a kidney disease gene that was compatible with living happily into adulthood.

I became obsessed with flipping coins, given that our chance of a fetus having the kidney disease gene was fifty-fifty. Heads, I have a baby, tails, I have an abortion. The coin would come up heads and I'd feel a little leap of joy. Then I'd toss it again and get tails. Then again: tails. The Project might lead to a grim sequence of events. I looked up images of twelve-week-old fetuses and saw that even though they were only 2½ inches long, they had little hands and legs. I wondered if I could go through with an abortion for this kidney disease even once, much less twice, despite what I'd told my husband. And even though he'd said reassuring words about not getting angry, I didn't know what it would do to my marriage if I terminated our child for having the same genetic makeup as him. I imagined that when this was over, once we had a healthy baby in our arms, our joy would melt away any painful memories. That didn't feel consoling. Thinking about all this made me feel sad and confused; sometimes this seemed like the stupidest idea ever. Especially because, as the months went by, I didn't get pregnant.

I had always assumed that I'd conceive quickly; I was in my early thirties, and healthy. My husband, however, was not. His blood tests showed that creatinine, a waste product of protein metabolism that's used to assess kidney function, was on the rise. Just a few months after we'd started trying to get pregnant, he came home one night and said that his nephrologist wanted him to start dialysis. I was shocked. This seemed so sudden. I shouldn't have been surprised; he was almost exactly the age his mother was when her kidneys failed.

We toured dialysis centers, trying to find the least bad one in the city, because he'd have to spend about twelve hours a week there. One Friday night after work, we checked out a place near Union Station, which had the advantage of being easily accessible by subway. Recliner-like chairs were lined up in one big room, with people in them hooked up to boxy dialysis contraptions and either dozing or watching television. The place looked grubby. The nurse walked over to a dialysis machine and started explaining how this device processed blood, seemingly indifferent to the fact that the blood flowing into it was coming from a middle-aged man she had not bothered to acknowledge. I abashedly said hello and asked him to please excuse us. The nurse laughed, exclaiming, "Oh, we're all like one big family here." I glanced at my husband's face and sensed the tension behind his polite, attentive expression. We had planned to see a movie after this visit. Instead, depressed, we just went home.

My husband dreaded dialysis and wanted to put it off as long as possible, even though he had started to feel sick. He frequently complained of nausea and abdominal pains. He was losing weight. I worried about him and tried to make his life as easy as possible; I woke up extra early every morning so that I could make him a bagged lunch and walk with him to work. I also grew suspicious that his illness might explain why months had gone by without our getting pregnant. I found medical articles that said things like "infertility is frequently a problem in male patients with chronic renal failure." No one involved in setting up The Project had ever suggested that kidney failure might impact our likelihood of conception. I went back to my gynecologist, who frowned and said, "I didn't realize his kidney disease was that advanced." She wrote out a prescription for my husband to have his sperm tested at a nearby fertility center. And because I asked him to, even though he had zero desire to have anything to do with an IVF clinic, he went there and ejaculated into a cup.

When we sat down with a fertility doctor to go over the results of the sperm analysis, he said my husband's sperm was borderline—not technically abnormal, but not super healthy, either. We asked what this might mean for The Project. Could we get pregnant the natural way? How long might it take? The whole conversation became a blur of numbers as the doctor scribbled success rate statistics for IVF and plain ol' sex on a piece of paper, trying to calculate what we might expect if we went either route. He was of the opinion that our chances of producing a child free of polycystic kidney disease in the immediate future seemed much higher if we did IVF with embryo testing and selection. A thirty-three-year-old woman like me with no history of infertility, he said, should respond incredibly well to ovary-stimulating drugs. There'd be lots of eggs, and that meant plenty of embryos to pick and choose from—an extremely important consideration, given that half of our embryos were likely to have the disease gene and would therefore be discarded. The doctor was so confident in our chances that he was willing to enroll us in the clinic's "shared risk" program. That meant we could pay a flat fee of $20,000 to do as many IVF cycles as necessary to get a baby. If it didn't work, we'd get our money back. (The drugs used to stimulate the ovaries weren't part of this deal. We'd have to buy them separately, and they were pricey.)

Hey, I was sold, despite the fact that speaking with this guy felt disconcertingly like talking to a car salesman. Making embryos and testing them is what I'd been wanting to do from the beginning, and it would put an end to the misery that was contemplating abortion. And my husband . . . well, my husband was sick. I don't know whether our numerous conversations had worn him down, or if our difficulty getting pregnant had changed his calculations about the acceptability of fertility medicine, or if he was just generally exhausted because he was in advanced renal failure and facing an imminent kidney transplant. For whatever reason, he agreed to try the IVF route; or rather, he agreed to let me try, because his role was simply to provide sperm once or twice while I let the doctor and nurses take over my body. Meanwhile, he was preoccupied with his own medical adventures. To get ready for dialysis, he had a vein and an artery in his arm surgically joined to create what's called a fistula, a giant single blood vessel that's strong enough to take repeated jabbing with a thick needle. The fistula grew huge and formed a bulge at his wrist. When we lay together in bed at night, with his arm wrapped around me, cradling my head on his chest, the fistula rested against my ear. I could hear the turbulent blood swirl around inside and feel a vibration through his skin that medical texts call the fistula's "thrill."

Nothing about this was thrilling. Even after the fistula was ready, my husband kept putting off dialysis. He was sleeping a lot; he looked gray and gaunt. I typed his lab test numbers into a website set up by the National Kidney Foundation, and it advised starting dialysis immediately. Even if people didn't feel critically ill, I read, renal failure can cause a high potassium level in the blood that can do permanent heart damage. I told a friend, "I would like him to start dialysis, you know, before the more severe symptoms show up—like coma."

To understand how to best support him, I joined him at one of his nephrology checkups. I asked the doctor why we couldn't just skip the dialysis and go straight to kidney transplantation. My husband had a younger brother who seemed to be free of this disease, and he'd volunteered to donate his kidney. I'd also be willing. The nephrologist explained that in his opinion, it was better to be on dialysis for a while. I forget the exact reasons he gave, but to me they didn't make any sense. The next day, sitting at work, I started doing research online. I soon learned that the relevant word was "preemptive." Doctors had started to do "preemptive" kidney transplants in people who weren't sick enough to require dialysis yet. It turned out—surprise!—that this approach produced better results. I emailed my husband, telling him, "I think your doctor needs to renew his subscription to the *New England Journal of Medicine*." I looked up the phone number of the kidney transplant coordinator at Johns Hopkins Hospital, 30 miles away in Baltimore, and called her to ask if they did preemptive transplants. The answer was: *of course*. Having a living donor, she said, was like having the "golden ticket" because if your volunteer wasn't a good match for you, they had ways of doing donor swaps with others in the same situation so that, in the end, everyone got a well-matched kidney. I asked what we had to do to meet with their transplant team as quickly as possible.

Our marriage now had become two simultaneous medical campaigns, one to produce a kid and one to procure a kidney. Both IVF and kidney transplantation involved an exhausting number of emails and phone calls and preparatory tests, with long delays when nothing seemed to move forward. One or the other of us was always going to a doctor's appointment or having blood drawn or calling the insurance company or hunting down lost medical records. Switching from fetal testing to embryo testing meant we had to send DNA samples to a different genetic lab, to build another custom test for the family mutation. My husband did a stress test for his heart, had his chest x-rayed, and was screened for HIV, hepatitis C, and tuberculosis. He was required to see a social worker to prove that he was emotionally stable enough to take immunosuppressant drugs reliably, without fail, for years. Meanwhile, I had dye squirted into my fallopian tubes and took "injection class" to learn how to jab myself with needles. I felt guilty every time I went into the IVF center and sat in the waiting room; I was sure that every woman there wanted a baby more than anything and would be *just fine* to be pregnant with one that had polycystic kidney disease. If they knew why I was there, I imagined, their contempt would be brutal.

Once the genetic test for embryos was ready, I got the go-ahead to begin IVF. I was excited. A year and a half earlier, I had walked into my gynecologist's office and told her that I wanted to try to have a baby. Now, within weeks, I might be pregnant—and with a kidney-disease-gene-free child, no less. I tried to remind myself that even after the clinic transferred one or two carefully selected embryos into my uterus, the bundle of cells might not implant and I could end up with a negative pregnancy test. But since I was so young and healthy, it didn't seem overly optimistic to hope for success with the first IVF attempt. A box of vials and syringes arrived in the mail (total cost: $3,700), along with a detailed schedule of which drugs to take and when. I'd stop in at the clinic repeatedly for blood work and sonograms of my ovaries, to measure my estrogen level and look for signs of developing eggs.

My first sonogram wasn't promising. It revealed four fluid-filled sacs on the surface of each ovary. Each of these sacs, or follicles, might contain a growing egg. A total of eight was hardly the enormous number that the doctor had expected me to produce. At our first appointment, he'd said that I'd probably make around twenty. And making an abundance of follicles was a key part of this whole embryo-testing plan. That's because some follicles might not contain eggs. Some eggs might not fertilize or might not develop into embryos.

Some embryos might not survive the poking and prodding needed to do the genetic testing. And the testing would likely reveal that some embryos had the kidney disease gene, so we wouldn't want to use them. The Project had come down to a numbers game, and our number of follicles (only eight!) didn't look good. But the doctor didn't seem totally discouraged. Maybe there were more follicles that he couldn't detect on the ultrasound. He told me to increase my dose of medication and we would see what happened.

What happened was nothing. My estrogen level stayed flat instead of rising, even though I was injecting myself with ever higher doses of the finest follicle-stimulating hormone money could buy. And instead of growing bigger, my follicles shrank. This was a completely unpredictable response, but the doctor's response to it was sadly predictable: he had the nurse call to ask if I was sure I was injecting the drug properly.

"Yeah, I'll *bet* he asked that," I said.

This was just so surprising, she said, given my age and my initial blood work.

I told her, "Look, I guess it's possible I am screwing up, but I really don't see how. You measure the stuff out, you see it in the syringe, you stick the needle into your skin, you push the plunger down. I mean, where is the liquid going if not into my body? I am fairly motivated to do this correctly. It's just not that hard."

She instructed me to take even more of the drug, as a last-ditch effort. That night I asked my husband, "Do you want to watch me pointlessly inject $727 into myself?"

Soon only three follicles were visible. The doctor said he was sorry I was having such a "slow response."

"Don't you mean 'no' response?" I asked.

He said we needed to call off this IVF cycle, and he'd think about what to try next. I stopped taking the drugs and waited to get my period.

I walked around in a daze. What had just happened? Sometimes I thought that my ovaries were smaller versions of how I liked to think of myself—unflappable professionals. I imagined the two of them commiserating like coworkers bitching about the latest random idiocy from upper management. "*More* follicle-stimulating hormone? What the hell? This is unbelievable! What are they thinking?" one ovary would say. The other would reply, "I don't know, sister, but listen, that shit is *fucked up*. We're supposed to make one egg per month, that's it, and that's exactly what we're going to do." This was a pleasant fantasy. Most of the time, however, I worried that my ovaries were somehow defective. Maybe that explained why I hadn't gotten pregnant before we turned to IVF. Maybe it wasn't just bad luck, or kidney-failure-addled sperm. My body, which had always been so hardy and dependable, was failing me.

After a couple of months, we tried IVF again with another drug regimen. This time the initial sonogram glimpsed seven follicles. And in response to the injections, my estrogen levels started rising. Although the upward trend was modest, I started to feel hopeful. Then the doctor scanned my ovaries and found that my follicles had shrunk. He could only see four.

He took me down the hall into a small room with thin Sheetrock walls and a couple of chairs. We sat down. He told me that he didn't think this was going to work. For whatever reason, my body wasn't producing enough eggs for us to have any reasonable belief that moving forward would result in our being able to make enough embryos to screen them for the kidney disease gene. We didn't even know that these four follicles would keep growing over the next few days or that each one would contain a mature egg. Egg retrieval was an invasive thing to do, and we might only get two or three eggs. These eggs might not fertilize, or if they did, the resulting embryos might end up with the kidney disease gene; given the fifty-fifty odds, we just didn't have the number of embryos we'd need to have a decent shot at successfully producing an embryo or two without it. He said it was a judgment call—some women would jump at any chance—but he didn't think we should go forward.

I agreed.

He had one other drug protocol we could try. Given how things had gone, though, he thought my ovaries just resisted stimulation and we'd have another bad result.

I agreed with that too.

We sat there together quietly for a moment. I sensed his relief that I was calm. I thought about all he must have witnessed in that bare, closet-like room, a private place likely designed for the delivery of bad news. So many times, a distraught woman must have sat in the chair I was sitting in, hearing that she'd never have a child that shared her genetic heritage. I knew that wasn't what he was telling me; my situation was different. I had no real reason to believe I couldn't get pregnant. Still, my reproductive efforts so far—both natural and artificial—had not inspired confidence. I was filled with fear.

"Look," I said to him, "what if I told you we didn't care anymore about screening for polycystic kidney disease? What if I said that I just wanted to have a baby, as soon as possible? What would you recommend? Continuing this IVF cycle, or what?"

"I would recommend going home," he said, "and having sex with your husband."

Try that for a year, he advised. If it didn't work, we could always come back and do IVF with donor eggs. With donor eggs, he said, he had no doubt that I'd get pregnant. And if we used another woman's eggs with my husband's sperm, he pointed out, then we'd surely be able to make tons of embryos. We'd have plenty! And that meant we would absolutely be able to test them for polycystic kidney disease!

My eugenic effort's mordant ending—the one I'd known was coming—had finally arrived.

The god of science and magic, which the ancient Egyptians called Thoth, would let me have my beloved husband's children and use eugenic tools to eliminate any possibility of starting a pregnancy that would have his kidney disease gene. But here was the price: our children couldn't inherit any of *my* genes. I'd have to throw my heart to the floor and let the soul-eater, Ammit, gobble up any chance that some small part of me could keep on living after I died.

God had probed my heart, and examined my kidneys, and repaid me according to my ways, with the fruit of my deeds.

The Project was over.

"I want our $20,000 back," I told the doctor, standing up to shake his hand and go.

I once made a compilation of songs about journalism that included "Front Page Story," by Neil Diamond. He sang that when someone's heart gets broken, that sad event won't appear in headlines or the newspapers, since broken hearts happen too often to be newsworthy. Similarly, the everyday defeats of reproductive medicine don't make compelling news stories. That means the public might overestimate what cutting-edge science can do.

Since the first genetic testing of embryos created through IVF, in 1990, this technology has allowed people to have children while avoiding hundreds of heritable conditions, including cystic fibrosis, sickle cell anemia, hemophilia, and muscular dystrophy. The press has covered grateful parents who say they've been able to have the healthy family they always wanted—along with a lot of hand-wringing about where this powerful technology might go in the future. But IVF plus genetic testing doesn't always produce a baby.

In fact, it *frequently* doesn't. One medical group in Paris, France, reported in a science journal that from 2011 to 2016, they tried this for 358 couples at risk of transmitting a single-gene disorder. "After several attempts, only 95 couples out of the 358 brought home a baby," the researchers wrote, admitting that, in other words, this strategy "had a failure rate of 73%."

That was unusually candid; certainly no one informed me of a track record like that before I made the decisions I did. But those researchers were only being so frank because they wanted to make a point about what could be done to increase the odds of success. Instead of just discarding embryos with an unwanted genetic condition, they argued, maybe doctors should try to genetically *fix* those embryos.

After all, one of the reasons for the poor results in that group's practice was that nearly a thousand embryos couldn't be used because they had tested positive for the couples' genetic disorders. If the researchers had been able to inject some kind of treatment into the embryos that could correct the troublesome genes, maybe more would-be parents would have had healthy babies. Or, as the scientists put it, "research on genome editing in human embryos should enable us to offer couples, one day, the possibility of rescuing their affected embryos."

They're not the only ones to make this argument, as technological advances have forced a reconsideration of reproductive possibilities long considered verboten. For decades, the engineering techniques that labs used to alter the genes of animals and plants were fairly crude and uncontrollable. Almost all scientists agreed that it would be irresponsible to apply them to human embryos, given the potential unknown effects on the child or future generations. In recent years, however, scientists have invented far more precise gene editing tools. What was once off-limits began to be discussed. Prominent geneticists and ethicists began to consider what initial applications might be worth the risks of trying to alter a human embryo's DNA.

Then, in 2018, all of this careful discourse got blown apart, almost as if someone had unilaterally detonated a nuclear weapon. A scientist from China, He Jiankui, announced that he had used an editing technique to try to disable a gene in human embryos, resulting in the birth of genetically altered twin babies. The gene in question, *CCR5*, is involved in HIV infection, and He apparently believed he was making the children resistant to this virus, which had infected their father. As news of his taboo-breaking act ricocheted around the world, horrified scientists quickly tried to distance themselves. "The project was largely carried out in secret, the medical necessity for inactivation of CCR5 in these infants is utterly unconvincing, the informed consent process appears highly questionable, and the possibility of damaging off-target effects has not been satisfactorily explored," said Francis Collins, a renowned geneticist and longtime head of the National Institutes of Health, in a statement released after the news broke.

It's worth noting the echoes of eugenics in this first genetic engineering of human babies. The scientific justification was shoddy. The parents involved reportedly weren't fully informed of what they were getting into. They also came from a disadvantaged community; HIV is stigmatized in China, and the expense of IVF there can put it completely out of reach for couples who might have seen participation in this experiment as their only way of having children. Given all that, it's no wonder that scientists who hoped to someday move forward with gene editing in human embryos wanted to portray this event as an aberration. In its aftermath, an international commission of experts was convened by the National Academy of Medicine, the National Academy of Sciences, and the United Kingdom's Royal Society, to consider what legitimate justifications there might be for attempting heritable human genetic modification, now and in the future.

This august group concluded that there wasn't enough evidence yet that any gene editing technique could be done safely enough in human embryos. But these experts didn't shut the door to future efforts. This path, they said in a press release, "should only be considered when prospective parents who are at known risk of transmitting a serious monogenic disease have no option or extremely poor options for having a biologically related child who is not genetically affected without the editing procedure, due to genetic circumstances or the combination of genetic circumstances and fertility issues."

When I read that, I felt amazed. Didn't I have a "combination of genetic circumstances and fertility issues" that was thwarting my efforts to have a child with my husband while avoiding passing on a serious single-gene disorder? In the quarter century that I'd been a science reporter covering this genetic stuff, the prospect of editing genes in human embryos had gone from being completely out of the question to something that was hitting rather close to home.

Starting to tinker with an embryo's genes would be very different from just selecting one embryo out of a group of them created by combining sperm and eggs. That's because the ability to rewrite bits of the human genetic code dramatically expands what might someday be possible. Once gene editing tools have been developed and perfected in human embryos to correct disease mutations, people might start to use those tools to achieve more questionable ends—and anyone who would doubt that has to contend with the fact that it basically already happened in China, when He tried to create babies that were immune to HIV. Gene editing in human embryos could move the world closer to the old eugenic goal of scientifically breeding "human thoroughbreds," assuming that researchers ever figure out any genetic changes that might improve cognition or other abilities. One could imagine a future dystopian world like the one in the sci-fi movie *Gattaca*, where genetically enhanced people known as "valids" discriminated against "in-valids," who were seen as inferior because they were conceived in the traditional, low-tech fashion.

That kind of fictional scenario can seem like outrageous fearmongering to pioneers of reproductive technologies. Embryologist Alan Handyside, whose group was the first to ever test IVF-created embryos for parents who wanted to avoid a genetic disease, looked back on the thirtieth anniversary of the achievement and recalled how the world absolutely freaked out when his team held a press conference in London, with his pregnant patients in attendance. Their unprecedented work in testing IVF-created embryos in order to select particular ones immediately "raised fears of 'designer babies' in which parents would use the technology to select desirable characteristics, such as hair or eye colour, or to prevent the inheritance of less serious conditions," Handyside wrote, adding, "In my experience, beyond some couples wanting to select for gender, there has been no evidence that this is the case and I have long argued that parents can be relied on to make their own informed decisions."

The phrase "designer babies" does convey an inappropriate air of frivolity, as if a parent wanted to create an especially cute child in the same way that he or she might go into a department store and overpay for fancy denim jeans. "I deeply despise the phrase 'designer children.' Fuck eye color, when genetic technology can't even give me children with whole brains," wrote one frustrated parent in an online forum set up by *The Atlantic* called "When Does Abortion Become Eugenics?" This particular parent carried an unknown gene for hydrocephalus and had aborted two fetuses with skulls filled with fluid before eventually having a healthy child. In 1999, in a *New Scientist* magazine article entitled "Designing a Dilemma," geneticist Mark Hughes expressed his dismay that it seemed like people couldn't ever talk about genetic testing of embryos without someone linking it to "designer babies" and eugenic breeding. "This is an incredibly complicated process that only people who are very desperate are going to bother with," Hughes said. "People are not going to come at this for trivial reasons."

I wrote that magazine article a year before I got married. And I believed what Hughes told me, then and now. But another part of me thinks that this sharp distinction that genetic researchers constantly make between "serious" and "trivial" concerns can obscure a lot of ethically murky ground. A decade after interviewing Hughes, when I personally chose to go through that "incredibly complicated process," I wasn't thinking of something as inconsequential as hair or eye color—but I also wasn't facing a genetic disorder bad enough to destroy a baby's brain. Maybe polycystic kidney disease wasn't "trivial," but it sure seemed like "not that big a deal" to my husband, who had the illness himself and saw no need to do anything extreme to prevent it. He could hardly be described as a person who was "very desperate."

And I wasn't either. Because it certainly had occurred to me, before I even started the IVF process, that I might make embryos and learn through genetic testing that we'd had really bad luck—that every single one of the embryos had the polycystic kidney disease mutation. If that was the case, I wouldn't want to throw the affected embryos out. I'd want to freeze them. Because for all I knew, those might be the only embryos I would ever create with my beloved husband, and I might well end up wanting to try to start a pregnancy with them. (Whether or not the fertility clinic would have gone along with implanting them is another matter.) If I knew I might potentially want to use such embryos, despite their having this disease gene, didn't that mean that my desire to try to pick and choose among the embryos was based on something "trivial"—or, at the very least, not do-or-die essential?

What's more, if my eugenic baby-making effort had happened in the not-too-distant future, at a time when it was possible for my doctor to offer me the option of continuing on with IVF and trying to fix the disease mutation inside any affected embryos instead of simply discarding them, I have to imagine that I probably would have wanted to go for it. It's not as if I had declined anything else that medical science had offered. Without even asking, I know what my husband's reaction would be to that idea: *Hell, no*. But would I have eventually worn him down on that, too? And what would it have done to my family relationships if my children eventually learned that I had rejiggered their DNA? Would they feel grateful to be able to live without a burdensome disease? Or would it feel like an unwelcome intrusion into their very nature, like they barely got a chance to exist before I started mucking around and trying to change them, to perfect them, without their knowledge or consent?

I feel fortunate that the technological times I lived in never forced me to make that kind of decision. As it was, I'd taken my poor husband on an expensive and time-consuming tour of the available eugenic possibilities, and we'd ended up right back where we started. The only difference was that I felt emotionally shattered by the experience and worried about our chances of ever having children together. My husband remained optimistic, believing that one way or another, we'd find a way to have a family. On the day when I told him that my ovaries had let us down once again, that I'd had that final conversation with the IVF doctor and had come to terms with The Project being over, he said, "That's pretty disheartening news. But it's also sort of liberating."

A month after that, we walked through the predawn darkness of Baltimore with his younger brother and their father, through the doors of the Johns Hopkins Hospital. The two brothers got prepped for surgery and wheeled into the operating room while their father and I hung around helplessly, wandering between the cafeteria and a crowded waiting area. We waited for hours. One of their surgeons eventually found us and said everything was perfect; the healthy kidney was implanted, our loved ones were stitched back up, and we could go see them on the recovery floor momentarily.

"Do you have any sense of how long it might take for the kidney to start functioning?" my father-in-law asked. To him, this was a natural question, because when his wife had her transplant, twenty-five years before, it took a long time for the cadaver kidney to kick into gear and start filtering her blood. The surgeon seemed startled, even amused by this question. He was a tall, jolly man, with red hair and a magnificent handlebar mustache, and he exclaimed excitedly, "Oh, the kidney is working already! Even as we were sewing all the vessels and everything up, we could see drops of urine coming out of its tube!" We marveled at my husband's creatinine level, which plummeted overnight as the transplanted kidney deftly cleaned his blood. This heroic organ must have been befuddled by its unprecedented trip outside of its body, and then horrified to get stitched into an abdomen that contained such inexplicable filth.

When my husband was declared well enough to go home, after about a week, we somewhat bitterly joked about sending out a notice to all our friends, similar to the joyful birth announcements that we received after their trips to the hospital. "It's a kidney!" we'd proclaim, with a photo of the organ plus its weight and length. I no longer had any interest in testing a fetus for my husband's kidney disease gene. I just wanted to get pregnant as quickly as possible, and I was afraid it wasn't going to happen; given all that had come before, there was now no way I'd have considered abortion to prevent this disease. The prescription information sheet for one of the sixteen drugs my husband was taking warned of DNA damage and birth defects. Fortunately, he only had to take that medication for about three months. We waited for three more months after that, so his sperm could turn over, before trying again.

And this time, right away, the pregnancy test was positive. We were so happy. Until a sonogram showed that the fetus had died. It felt like a punishment. It felt as if all the gods I don't believe in wanted to underline *exactly* how wrong I'd been to consider ending a pregnancy because of a treatable genetic disease, one that my husband lived with every day. For various reasons, I wanted my doctor to remove the dead fetus rather than waiting for it to be expelled naturally, and as I went through the surgical prep and met with the anesthesiologist, I kept thinking about how, at one time, I'd been planning to take every one of these steps to end a viable but affected pregnancy. I had taken my body's vitality for granted; I had assumed I could get knocked up again and again. I cringed at my hubris. I had been painfully naive.

The next time I got a positive pregnancy test, I felt uncertain for months. I wanted to give birth as privately as I could, at home, with certified nurse midwives who mostly wouldn't touch me or interfere. The day I went into labor, I paced around my house and spent a lot of time in a rented inflatable tub full of hot water. I had planned to use the tub only for pain relief, but then the baby's head came out while I was underwater. Afraid it was drowning, I thrashed around in a panic. The midwives told me to stay in the water and tried to reassure me that the baby wouldn't try to breathe until it hit the air. Their logic couldn't calm me, and when the next contraction hit, I pushed frantically, as hard as I could, not caring if I ripped open. A warm, smooth, pink body floated up. I grabbed it beneath its arms and got its head out of the water. I was looking right into a small, wrinkled face. It gave a cough and opened its eyes a slit, squinting to the left and to the right. My own eyes, captured in a photograph, were so wide that the circle of the dark pupil was entirely surrounded by white. I exclaimed, "It's—it's a baby!"

Three years later, my daughter was born. I held on to the side of the rental tub, which was set up in our guest room. This time I was standing up as the baby emerged, and the midwives quickly untangled her from the umbilical cord and shoved her onto my chest—thwack! The heavy, wet weight came as a shock. I wrapped my arms around the baby and stumbled to the bed. The baby was huge, over 10 pounds, and smeared with blood and waxy white vernix. The skin looked alarmingly blue. "Is it OK? Is it OK?" I asked, distraught, as the midwife assured me that the baby was fine. My husband joined me on the bed, and the midwife climbed on, too, to reach the baby. Then the bed collapsed. I had to shuffle down the hallway to my own bedroom, the umbilical cord connecting me to the human being wrapped in a blanket in my arms.

All of that feels so long ago. Sometimes—not often—I wonder if either of my children has the polycystic kidney disease gene. As I write this, I don't know. There's no medical intervention that could help at this early stage that would require knowing. My children are now full-fledged human beings with a right to genetic privacy; I would no more root around in their genetic code than I would break the heart-shaped lock on my daughter's pink diary. The fact that kidney disease runs in their family, however, is not a secret from them or the world. My mother-in-law's obituary discussed her father's death from renal failure as well as her own kidney transplant, and my brother-in-law has publicly talked about donating a kidney to my husband. There's a 25 percent chance that both my kids got the disease gene. There's a 25 percent chance that neither of them has it. There's a 50 percent chance that one of them has it and the other does not; if that's the case, our family would have a sufficient number of healthy kidneys to go around, assuming my children remain close enough that one would be willing to share with the other.

Before too long, in their teenage years, they will probably want to find out their genetic status. That's because medical treatments did advance, just as my husband predicted. Eighteen years after we got married, and eight years after my first child was born, the Food and Drug Administration approved the first drug ever for polycystic kidney disease. It slows down the progressive formation of cysts. And while it's not a perfect medicine—it's incredibly expensive and requires drinking large amounts of water—it can delay organ failure, potentially for years. Other pharmaceuticals are in clinical trials, so more medicines may be coming.

These children's great-grandfather died from this disease, back when doctors had nothing to offer. Their grandmother ultimately died because of it, too—although dialysis and a difficult kidney transplant from a cadaver gave her twenty extra years. Their father, my husband, has had a much easier time. He never had to endure dialysis, and his transplant has worked flawlessly; the drugs he takes have had no burdensome side effects. If one or both of my kids did get stuck with this crappy disease, their experience will surely be even better. Like any parent, I hate to think of them going through any suffering or uncertainty. But given the realities of existence as a mortal human being, not even the boldest proponents of eugenics ever promised to completely do away with that.

One recent rainy afternoon, looking for something to do with my children, I decided that we should go check out the National Museum of Health and Medicine. That's the museum that had the polycystic kidney on display, the one my husband and I had stumbled upon way back when. The museum's collections had all been moved into a snazzy new building, a place I'd never been, and I imagined that the kids might enjoy the chance to gawk at medical oddities like the bullet that killed Abraham Lincoln. They were skeptical of this outing, so I told them in the parking lot that if they didn't like the museum, we could leave.

We walked in together and immediately saw a towering human skeleton in a glass case. "Is that a real skeleton?" my son wanted to know. "Um, yeah," I said. The next case had anatomical specimens, such as a severed leg with elephantiasis. Down at the bottom was a row of jars that contained deformed fetuses. One of them had a squashed head and bulging eyes. Another was a pair of near full-term conjoined twins, fused at the chest, with sweet baby faces that were looking at each other. "Let's go home," my son urged. My three-year-old daughter stared at the malformed babies, displayed at her eye level, transfixed. She had questions. Why were they connected like that? Why were they in jars? Were these real babies?

I calmly explained that when babies start growing, there can be problems with how different parts form, so sometimes the baby dies. Doctors can study the dead baby to help them figure out how parts normally grow, I blandly continued, careful to hide my growing sense of dread. I had seen fetal specimens like this before, sure. Standing there next to my young daughter, though, I no longer felt detached. Whose fetuses were these? How would I feel if the fetus that had died inside me was put on display? I looked around wildly, searching for some less disturbing exhibit, and spotted a tall column of liquid. At the top was a brain. Below it, a gangly spinal cord hung down like an enormous millipede. "Wow, a real brain!" I said, trying to distract my daughter from the fetuses, while my son tugged on my arm, pulling me toward the exit. "Let's go, let's go, let's go," he repeated, dragging me past a row of child skeletons arranged by height, starting with one the size of an infant and progressing to ones about the same size as my children. Why had I brought them to this place? Was I insane?

We spent approximately three minutes in the museum. My old nemesis, the polycystic kidney, must have been in there, but I was so focused on my children, I forgot to even look.

Once we got outside, into the cool misty air, everyone relaxed. My daughter reached up and took my hand as we stomped through puddles. "I guess those babies were just too weird to live, right?" she said, cheerfully.

I helped her into her car seat and said that some conjoined twins do just fine. I told her about Chang and Eng Bunker, how they lived for a long time, how they had a farm and got married to two sisters, and how they each had lots of children. "Can we please stop talking about this?" said my son, from the back seat.

"Aren't you glad we didn't come out connected like that?" my daughter asked him. He shrieked with laughter, shouting, "That's not possible!" I reminded her that her brother had not been in my uterus with her. When she was a fetus inside of me, I said, he had already been born and was about the same age that she was now.

I watched as my daughter tried to absorb this information. I could tell that it was hard for her to fully grasp the idea of a time before her own existence, to understand that the past had truly happened and had brought us to this moment, sitting together in the car in the rain. Some instinct told her that she was on the verge of understanding something, that what she had just seen held some kind of meaning, and it had something to do with her. "When I was born," she persisted, "were you glad that I had all the right parts?" And I told her that yes, I was glad. When she was born, she was perfect. And I was very, very glad.

Acknowledgments

Thanks to my ol' pal and book accoucheur Tim Kreider; agent Meg Thompson; editor Matt Weiland and assistant editor Huneeya Siddiqui at W. W. Norton; copy editor Janet Greenblatt, project editor Susan Sanfrey, art director Sarahmay Wilkinson, publicist Erin Sinesky Lovett, and marketer Steve Colca; wise advisers Michelle Nijhuis, Emma Marris, and Malaka Gharib; early readers Ann Finkbeiner, Barton Kraff, Meg Guroff, Aaron Long, Adrian Cho, and Laura Helmuth. And my family, of course, especially David and my two dear kiddos—how I love you so.

Selected Sources

Bluestein, Howard B. *Tornado Alley: Monster Storms of the Great Plains*. New York: Oxford University Press, 2006.

Buhler, Andrea Jean. "Fanciful but Not Forgotten: A Historical Examination of the Study of the Flea, 1840–1930." MA thesis, University of Wisconsin-Milwaukee, 2017. https://dc.uwm.edu/etd/1590.

Burke, John G. *Cosmic Debris: Meteorites in History*. Berkeley: University of California Press, 1986.

Fujita, T. Theodore. *Memoirs of an Effort to Unlock the Mystery of Severe Storms, During the 50 Years, 1942–1992*. Chicago: Wind Research Laboratory, Department of Geophysical Sciences, University of Chicago, 1992.

Herdeiro, Carlos A. R., and José P. S. Lemos. "The Black Hole Fifty Years After: Genesis of the Name." *ArXiv.org*, 12 Dec. 2019, https://arxiv.org/abs/1811.06587.

Larsen, Jon. *On the Trail of Stardust: The Guide to Finding Micrometeorites: Tools, Techniques, and Identification*. Beverly, MA: Voyageur Press, an Imprint of The Quarto Group, 2019.

Lombardo, Paul A. *A Century of Eugenics in America: From the Indiana Experiment to the Human Genome Era*. Bloomington: Indiana University Press, 2011.

Lombardo, Paul A. *Three Generations, No Imbeciles: Eugenics, the Supreme Court, and Buck v. Bell*. Baltimore: Johns Hopkins University Press, 2022.

MacLay, W. S., E. Guttmann, and W. Mayer-Gross. "Spontaneous Drawings as an Approach to Some Problems of Psychopathology." *Proceedings of the Royal Society of Medicine* 31, no. 11 (Sept. 1938): 1337–1350.

Nininger, Harvey Harlow. *Find a Falling Star* (Introduction by Fred L. Whipple). New York: Paul S. Eriksson Publisher, 1972.

Robertson, John A. *Children of Choice: Freedom and the New Reproductive Technologies*. Princeton, NJ: Princeton University Press, 1996.

Stern, Alexandra Minna. *Telling Genes: The Story of Genetic Counseling in America*. Baltimore: Johns Hopkins University Press, 2012.

von Petzinger, Genevieve. *First Signs: Unlocking the Mysteries of the World's Oldest Symbols*. New York: Atria Paperback, 2017.

Wallace, Anthony F. C. *Tornado in Worcester: An Exploratory Study of Individual and Community Behavior in an Extreme Situation*. Washington, DC: National Academy of Sciences, National Research Council, 1956.

Watson, Ben. "Oodles of Doodles? Doodling Behaviour and Its Implications for Understanding Palaeoarts." *Rock Art Research* 25, no. 1 (May 2008): 35–60.